images,
images,
images

Written for Kodak by
Michael F. Kenny and Raymond F. Schmitt

© Eastman Kodak Company, 1979
First Edition, First Printing
Standard Book Number 0-87985-222-4
Library of Congress Catalog Card Number 78-74981

1

There are few eminent experts of long standing in the business of multi-image production; the field is too freshly planted to have produced a harvest of "recognized authorities." Instead, the territory is populated by explorers and experimenters—people who have stretched their imaginations beyond the confines of a single, small screen; who have grafted image to image to even more images in synthesizing a new visual medium; who have developed the sophisticated equipment and techniques that have transformed the process of presenting visual information; who have risked reputations and profits in the pursuit of innovation and who, in the process, have taught us all how multi-image presentations are produced. The efforts of these pioneers to extend the boundaries of visual communication created the need—and the content—for this book.

During our research and writing, we talked with a number of these producers, manufacturers, and production specialists; they generously shared their lessons and experiences with us, and for that we want to give thanks.

Chermayeff and Dorothy Bearak; Paul Condylis, John Condylis, and Christian Mendenhall; Laurence Deutsch; Thomas Foley; Robert Kirchgessner and Barbara Liber; Richard Shipps and Vincent Bonacci; and David Wynne.

● Next, to the equipment manufacturers whose products are mentioned or pictured on these pages, for their technical assistance and guidance and for their patience as we examined and photographed their equipment.

● The previous listings, of course, are incomplete. We have also learned about multi-image production and presentation from a great many people. Some we may have overlooked inadvertently; for that we apologize. In most cases, however, we just don't know the names of the hundreds of producers, directors, photographers, writers, designers, sound specialists, and equipment engineers and technicians who have taught us through their work. They too deserve credit for the ideas and suggestions that appear on the following pages.

2

Table of Contents

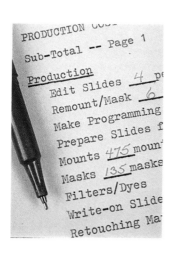

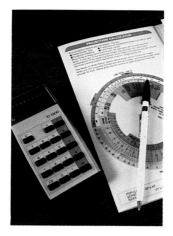

SECTION III
PRODUCTION 111

Introduction

Just What Are You Getting Yourself Into?

You're probably reading this book for one of two main reasons.

It could be you're a novice in multi-media or multi-image production. You've read about the techniques in audiovisual magazines, maybe you've even seen several impressive presentations, and now you'd like to use them in your communications program. But you don't know how to get started. In fact, you may not even be sure you **want** to get started. When you read the accolades in magazine articles you also read even lengthier explanations about the hardware, the time, and the cost involved. And if you looked behind the scenes at a multi-image presentation, you saw dozens of projectors, a sophisticated tape deck, and a unit that looked like a computer terminal — all of which were connected by what seemed like miles of wire. If you asked any questions, you might have heard such terms as digital information, cue links, freezes, supers, high-speed sequences, memory dumps, projector banks, etc. If you asked about the cost of all the gear, you may have heard figures that approach your audiovisual budget for the year. And now you're reading this book and wondering — just what am I getting myself into?

Or, maybe you're reading this book because you're an experienced multi-media or multi-image producer. The articles in the audiovisual magazines featured your shows. And you were the person behind the scenes, explaining to the uninitiated how you program a number of different dissolve rates and how you give tray update commands and how you build sequential loops for animation. You hooked up the projectors, dissolve modules, tape deck, and programmer; the maze of wires is part of your job. And you've paid the bills; you know a multi-image show isn't a casual undertaking. You're a professional. But you also know there's still a lot to learn. You've kept up with the technology of multi-image production, and you know it's changing every year. Programmers are becoming faster and more sophisticated. Some of the control work load is being shifted to advanced projector control modules. Programmers are controlling other programmers. As you consider the trends shaping multi-image production, you may be wondering—just what am I getting myself into?

The answers to that question—whether you're a novice or an experienced producer—are in this book. That may sound like an exaggerated claim: a book for both beginners and professionals that will appeal equally to both. But that's exactly what it is. In the following pages you'll find the information you'll have to consider before deciding to produce a multi-image presentation. You'll learn to decide if multi-image is appropriate for your audience, your purpose, and your occasion. You'll learn what sort of staff, equipment, time, and budget you'll need to produce a professional multi-image presentation. You'll also learn what to look for when evaluating suppliers—from processing laboratories to professional producers.

If you decide to go ahead with your presentation, you'll also learn **how** to implement your decision. You'll learn how to plan, how to produce, and how to present. More specifically, you'll learn how to budget, how to direct, and how to stage and control multi-image presentations. To make this book helpful to both beginner and professional, it's been written so that it can be read selectively—you can read what you need to read.

If you're a **decision-maker**—a person who must approve the creation of a multi-image presentation without actually becoming involved in its production—you can limit most of your reading to Section One: The Initial Decision. There you'll find information about when, where, and why you should use multi-image presentations and how you go about organizing and budgeting for them.

If, on the other hand, you're a **communicator**—the person who must plan for and assemble all the parts of a multi-image presentation—you'll want to read quite a bit more. The three remaining sections of the book—on planning, production, and presentation—plus the appendix are aimed at you.

Sophisticated programming equipment controlling 30 projectors allowed producers at Photosynthesis Inc. to create a sequence of dancing couples for a presentation for Audio Visual Laboratories, Inc.

International exhibits and World Fairs were early proving grounds for multi-image presentations such as this Cavalcade of Color in the Kodak Building at the 1939 New York World's Fair.

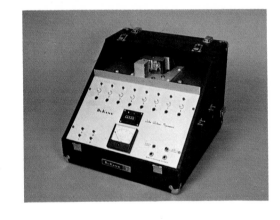

DuKane Corporation's Custom Electronic Programmer, the first commercially produced multi-projector programmer, offered a capacity of eight channels.

But even here you can read selectively because the book is structured to keep basic material and more detailed supporting material separate. For example, the chapter entitled Recording Sound, starting on page 150, is devoted to just that. The basic material in this section includes information on recording techniques used in the studio and on location. Its purpose is to teach you to make good audio tracks—or to enable you to work with a professional engineer who can. But you'll also find more detailed information—such as how and when to use multi-track recording techniques, the advantages and disadvantages of single- and double-system sound recording for film, and tips and techniques professionals use when editing sound tracks. This separation of basic and advanced information results in a book that's easy to read, understand, and use, regardless of your level of experience.

You'll also discover examples throughout the book. Some are case histories; a number of professional producers and users of multi-image presentations have shared their experiences. The techniques and suggestions of these professionals are passed along. Other examples in the book are hypothetical; they represent situations commonly faced by multi-image producers. The methods of handling these situations are explained in step-by-step guides.

Where appropriate, the book also contains checklists and planning guides to help you organize your own projects.

In short, the book is meant to be a tool you can use again and again.

Before you delve into the information in this book, it might be of value to consider its title, especially the focus on the term "programmed" in relation to multi-image productions. The term is a reflection of the

sweeping technological changes taking place in the audiovisual industry — especially in that segment of it dealing with the use of 2 x 2-inch (50 x 50 mm*) slide transparencies. Multi-image productions aren't new, of course; they were used at world's fairs and industrial expositions before the turn of the century. But for the most part, still-frame presentations produced for industry, education, and government — as recently as 15 years ago — were limited to filmstrips or one-projector slide shows. The subsequent development of more advanced slide projectors and automatic dissolve controls made it possible to produce constant illumination presentations with two projectors. These advances, coupled with advances in systems for synchronizing slide changes with a recorded sound track, brought a new level of professionalism to slide presentations. And, at the same time, these developments served to whet the creative appetites of more innovative producers who recognized the potential for still further sophistication.

Some of the more technically oriented producers began to build customized gear to serve their needs. Then, the major equipment manufacturers introduced the first wave of commercially available programmers. Multiple-projector control had arrived for the masses, and with it the impressive sounding term "multi-media."

By the early 1970s a second generation of programmers emerged, offering a variety of dissolve rates, fast cuts, alternate flashing, and even programmed timing cues. Then, as the '70s reached their halfway point, the next major breakthrough occurred. Electronic programmers revolutionized audiovisual production. Not only could more sophisticated effects be created on the screen, but now they could be produced in a fraction of the time.

Producers and manufacturers created customized programming devices to meet the requirements of early multi-image productions.

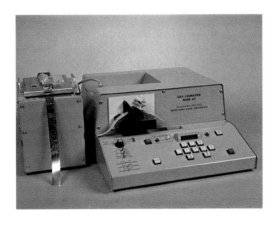

United Audio Visual Corporation's Cuemaster Mark 60 was one of the first mass-produced programmers to feature high-speed tape advance.

---

*For ease in reading, metric conversions are given once per dimension.

9

What's In A Name?

One organization that has tried to give precise definitions to the explosion of techniques in audiovisual production is the Association for Multi-Image. In its book, **The Art of Multi-Image,** multi-media is defined as "the coordinated use of more than one medium for the presentation of information." Multi-image is defined as "a presentation that generally uses the 'ideal' number of three projected images (but is not necessarily limited to three), an electronic programmer, and sound reproducing equipment."

A somewhat different set of definitions comes from Tom Hope, publisher of **Hope Reports,** a continuing examination of the audiovisual marketplace. Hope defines multi-image as "two or more pictures projected on one large screen or several screens at the same time." He calls multimedia (his preferred spelling) "two or more independent media in the same package or kit (program)." Multimedia, he says, is "software"; multi-image is "a presentation using hardware."

Still another definition comes from Gene Balsley, an audiovisual professional who comments on developments in the field. Writing in the January 1978 issue of **Photomethods,** he defined multi-media as "a slide show augmented, usually, with film, videobeams, live speakers, actors, and even life-size puppets." The element that transforms a basic slide presentation into a multi-media presentation, he added, is automatic control. That's three authorities and three different sets of definitions. And if you were to pick up other books or magazines dealing with the subject, chances are good you'd add other definitions to the list.

Producers have been searching long and hard for a term that would accurately describe their productions. "Multi-media" was popular for a while, but with the increased use of sophisticated equipment and techniques, producers and users began to coin other terms to describe their work. Multi-projector. Multi-screen. Multi-image. Multi-vision.

Trying to define these terms to everyone's satisfaction has proven to be more difficult than coining them. In fact, even writers on the subject can't agree on how to spell the terms. Is it multi-media or multimedia? And some can't agree on the terms themselves. Is the proper designation for a sound-slide presentation "multi-media" or the more recently coined "bi-media"?

Admittedly it's hard to find one term that adequately describes the growing number of options available to the audiovisual producer. Much of the difficulty, however, seems to center around the tendency to lump together definitions that deal with hardware (multi-projector, multi-screen), media (multi-media), and the visual product (multi-image, multi-vision). If these categories can be kept distinct during discussions of audiovisual productions, much of the confusion can be eliminated.

A second reason why the existing definitions have failed to adequately describe the state of current audiovisual art is the tendency of people to cling to word forms beginning with the prefix "multi." It seems to have derived from the thinking that if one of anything is good, then more—or multi—must be better.

But it is not the simple addition of projectors or screens or images or even media that is significant. After all, that was done years ago. The only difference between the early spectaculars and the presentations produced today is the sophistication and reliability of the control methods. It isn't so much that creative visual frontiers have been explored as it is that equipment has been developed that makes those explorations possible. That's what is significant about the new generation of still-frame audiovisual productions. But this is one point that most definitions fail to emphasize.

So, to repeat an earlier question: Just what are you getting yourself into? The answer of this book is: **You're dealing with the programmed presentation of visual images to convey information and create an impression.** The key word is "programmed." Programming enables you to manipulate images: to dissolve from one image to another at any of a number of rates; to animate; to integrate slides, motion pictures, light displays, and other special effects into an automatically controlled, consistently repeatable production.

In short, programming is the basic tool that makes it all possible; the projectors, or screens, or images, or media are simply available options.

The sophistication of your programming equipment, more than any other factor, will answer the question of what you are getting yourself into. Your imagination, talent, time, and budget will of course greatly influence what you can do; the greater your resources, the greater your potential for achievement. But if there is one limiting — and liberating — factor in determining the scope of your proposed presentation, it is programming equipment. Of course, the greater your programming ability, the greater your own flexibility to create.

On the simplest end of the scale are presentations using mechanical or magnetic cues to advance slides in one or a number of projectors, or one or more simple dissolve control units. The most common application for this type of programming involves two-projector dissolve presentations, for which the narration, music, and sound effects are recorded on one channel of an audiotape and the electronic cues on another channel. During a performance, a tape playback unit amplifies the sound signal and transmits it through a speaker system. At the same time, the magnetic cues are detected by a dissolve unit that controls the operation of the slide projectors.

A simple two-projector dissolve setup.

This type of electronically cued presentation is relatively easy to produce and is extremely reliable. And although it is comparatively simple in terms of programming sophistication, it need not be simple in terms of the visual effects it can achieve. The use of split screens, slide masks, and such photographic techniques as burn-ins, posterization, and sandwiching (all of which will be discussed in this book) can result in visually exciting presentations.

On the other end of the scale are shows that use multi-channeled paper-tape programmers or electronic micro-computers with built-in memories to control the operation of dozens of slide projectors **and** motion picture projectors **and** special lighting effects **and** moving displays

Taking advantage of the creative freedom offered by sophisticated equipment, producers are developing special screen formats for their productions.

. . . **and** you name it. These presentations are complicated even further by the fact that there is virtually no compatibility from one manufacturer's equipment to another's. There are, in fact, some instances where this incompatibility exists between the different models of programmers built by the same manufacturer. But the increased complexity of production is generally well worth the effort when the job is done. There appears to be no limit to the innovative variety of effects that can be achieved.

This sort of flexibility and creative freedom has been attracting growing numbers of producers and organizations to the use of programmed presentations.

- Tourists, with increasing frequency, are being drawn to such programmed presentations as **Jubilee,** in New Orleans (see box), **The Chicago Odyssey, Where's Boston?,** and **The New York Experience.** At Disneyland and Disney World, the art of programmed presentations has attracted — and continues to attract — millions of visitors.

- In education, programmed presentations are used in classrooms from elementary school through college. Many universities even offer courses in multi-media and multi-image production, a reflection of the growing interest in the medium.

- In business, programmed presentations have long been used as advertising and sales promotion vehicles. More and more training departments are beginning to use advanced programming capabilities to produce training programs that literally "present" themselves.

The reason for this increased use of programmed presentations grows out of the medium's flexibility. The ability to manipulate images provides communications benefits that most single-image media — slides, films, and video — can't give.

Images fill the screen in a multi-media, multi-image presentation called **Jubilee**, a 23-minute show focusing on the history, culture and tourist attractions of New Orleans.

One For The Money

During the past few years, a number of privately funded programmed presentations have been produced as tourist-oriented information and entertainment vehicles in major cities across the country. One of the most recent additions to the list is entitled **Jubilee**. It's located on Jackson Square, in the much-visited French Quarter of New Orleans, and focuses on the history, architecture, music, and culture of the Mississippi Delta region.

Planned and produced by David Wynne of Multi-Media Entertainment Corporation of America, **Jubilee** is heavily promoted as a tourist attraction by local travel agencies and tour guides. Unlike programmed presentations produced for use in business, industrial or educational environments, **Jubilee** and other similar presentations are produced as profit-oriented ventures; visitors are charged an admission. And what they are buying is sophistication.

Jubilee is unique in that the visual performance is not confined to a single theatrical arena. Visitors proceed through a series of five separate presentations—consisting of slides, super 8 and 16 mm motion pictures,

lighting effects, and holographic displays—that lead them to the main presentation theatre. There they view a 23-minute multi-image presentation projected by 15 slide projectors and a 16 mm motion picture projector controlled—along with moving fish-nets, Mardi-gras masks, and rotating ballroom mirrors—by a computerized programmer.

One factor that gives **Jubilee** its high degree of sophistication is Wynne's use of one programmer to control other programmers. A central programmer containing a continuous-loop paper tape controls projectors, light displays, screens, and the programmer used in the main presentation theatre. It also controls the crowds visiting the performance. By starting and stopping projectors, brightening and dimming lights, opening and closing screens, even opening doors, this central programmer in effect sets the pace for the visitors' progress through the 46-minute performance.

13

At the top of this list of benefits is the capability of the programmed presentation to project multiple viewpoints of the same subject—and to project them simultaneously. This characteristic of the programmed, multi-image presentation enables viewers to see objects, events, or settings in perspectives impossible through normal human perception. The viewers can see not only the panorama, but also close-ups of objects within the panorama. They can see the whole and the parts. They can see the top and the bottom, the front and back, and the sides of an object. They can see the beginning, middle, and end stages of a sequence—all at the same time. They can even see an activity both in motion and in still-frame analysis—at the same time. In short, programmed multi-image presentations give viewers not only details, but also the context into which those details fit.

This capability for multiple viewpoints produces a second benefit: more sophisticated organization of visual material. Using multiple images and programming, a producer can take a unit of information—such as the features of a new piece of office equipment—and present them to his or her viewers both in whole and in part. The viewer may first see a full shot of the equipment, then close-ups of the various features. Each close-up can appear as it is discussed; its position on the screen could correspond to its actual placement on the machine. This organizing and relating of material not only makes information more meaningful, it also helps improve retention.

Because they enhance perception and help organize and relate material, multi-image presentations allow producers to compress the time needed to create impressions. In single-image presentations, impressions are created through a process of addition: Images are added to images, one at a time, until a thought is conveyed or an impression is created. But with multi-image presentations, impressions are created by multiplying images. The impressions are not built so much as revealed. This is possible because a multi-image presentation in some ways duplicates—and in some ways surpasses—the process of human perception. We normally see in panoramas, our

(above) The images used to fill this free-hanging screen format—also pictured on page 12—seem to float in air. (right) Dozens of images fill the screen in this rear-projection presentation produced for ZDF at the Second German TV Programme Radio and TV Exhibition, West Berlin.

eyes sweeping the space before us, until something of interest arrests our attention. Then we narrow our focus and concentrate on that object. If that object is of great interest to us, we may narrow our focus even more.

This process of selectively "seeing" the world around us takes place rapidly. Our eyes may focus on dozens of objects in a matter of seconds. From these individual focusings our mind begins to build an overall impression.

Multi-image presentations duplicate this process. They can show us the entire scene, the full cycle of activity, the whole. At the same time they can show us the individual points of interest within the whole. The impression is created quickly. The process is not one of addition but of multiplication; the images don't "add up" to something; they reveal their meaning, full-blown.

Programmed presentations have the power to compress time and alter space. If produced effectively, they also create interest and excitement. Interest is built when aural and visual information is conveyed to a viewer at a pace that will hold his or her attention. (However, the use of too many visuals, too rapid a presentation of visuals, or visuals that are unrelated, can lead to information overload, in which the viewer is left confused, distracted, or annoyed. The question of pacing will be dealt with later in the book.) Excitement is created when the manipulation of images is artistic, choreographed to involve viewers in the presentation. Such choreography involves the use of color, movement, composition and sound—all subjects to be covered in the book.

But enough preliminaries. You now know what you're getting yourself into. You're going to be creating multiple impressions. To do this you may use slides, film, audiotape, holographic displays, lights, music, and even actors, dancers, or narrators; you'll project your images from three projectors or three dozen projectors; your images will be seen on one screen or a room built of screens; and you'll control all of this activity with a reel of audiotape, a spool of punched paper tape, or a computerized electronic programmer. But no matter what route you take, no matter what hardware and software you use, no matter the simplicity or complexity of your presentation, you're getting yourself into a medium that promises excitement, hard work and creative satisfaction as you fulfill your communication objectives.

Entrance at the original site of the 40-projector presentation **Where's Boston?,** produced by Cambridge Seven Associates as a Bicentennial salute to the city of Boston. The presentation is now located in the city's historic Faneuil Hall area.

This section is intended primarily for decision-makers. Its purpose is to help you plan the overall production of a multi-image presentation—to establish the objectives, goals, priorities, budget, schedule, and resources that will shape the completed production. Once spelled out, these decisions will help your production team gather, sort, and assemble the audio and visual elements of the show.

SECTION I
The Initial Decision

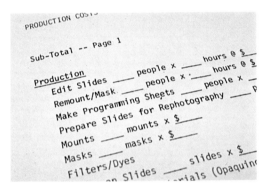

A precise statement of communications objectives helps determine how, when and where a program will be produced and presented. This 75-screen, 150-projector "information board" welcomed visitors to an exhibit for ZDF and ARD radio and television in West Berlin.

EVALUATING AUDIENCE, OCCASION, AND REQUIREMENTS

Before you launch your first—or next—multi-image presentation, ask yourself an important question: What are your communications objectives? Sound too basic or simplistic? It's not. People often jump into major productions without first analyzing their objectives and making certain that multi-image is the answer to their needs. The resulting presentation usually lacks impact and effectiveness. So, you should start your planning with a purpose. That purpose should be framed in a communications objective—a statement of the effect you want your presentation to have on an audience. Ask yourself: What do you want your audience to **do** after seeing your presentation?

For a customer-directed sales presentation, your communications objective might be "to motivate prospects to ask for an equipment demonstration." It also could be "to prompt the purchase" of the new equipment. For a presentation directed at your sales staff, the objective might be "to build enthusiasm for the new product." Communications objectives for a training program or a community relations program should reflect similar specific aims.

Writing a communications objective isn't a matter of simply describing your presentation's focus or theme. To do that is to confuse a statement of content with a statement of audience reaction. To say that you want "to explain the features and benefits" of a new product isn't a communications objective, unless, of course, your aim is to prepare the audience for a test. Such a statement merely outlines the content to be used in the presentation.

In stating your communications objective, you should frame your goal in terms of some change in human behavior. Your communications objective should always include such statements as "to ask," "to call," "to order," "to write," "to learn and use (a new procedure)," "to become excited about," etc. You might actually discover that your objective isn't so singular. There might well be—and usually are—two, three, or four goals. That's fine, but get them on paper and assign priorities to them, being sure that they're mutually compatible.

Now you're ready for some other, more specific questions:

- Can your communications objectives be accomplished with a multi-image presentation?
- Are the audience and occasion suitable for a multi-image presentation?
- Do you have the time and money to complete the production?
- Does your organization have the staff— or can you hire outside producers— to handle the production?
- Do you have the equipment, supplies, and space necessary to produce and present the show?

Asking these questions is just another way of reminding yourself to plan carefully before you begin a multi-image production. The importance of initial planning can't be stressed too much. It is, without exaggeration, a factor just as critical as the writing, photography, or programming in bringing a presentation to life. That's because planning helps you pinpoint your audience and the most effective approach to take in reaching that audience. Planning helps you set production goals and priorities. Planning helps you establish an adequate budget for multi-image production—or it helps you modify the scope of your presentation to an existing budget. Planning helps you establish schedules and evaluate staffing requirements. In short, planning helps you spot both the problems and the opportunities that will face you during the course of production; it also allows you to solve the problems before they arise and to take advantage of the opportunities as they appear.

Being in a position to eliminate problems and maximize opportunities early in your planning will assure you a highly satisfying return on the investment you make in the time and talent needed for multi-image production. And this investment can be sizable. The client may be spending from $10,000 to more than $250,000 for a multi-image presentation, so you want results—significant results. You and your client want the message to have impact; you both want it to be remembered, and to be talked about. Quite simply, you want to influence audience thinking and behavior. The way to do that is to take advantage of every opportunity your presentation offers. And the only way to create and profit from opportunities is to plan for them.

This sort of planning, of course, requires a great deal of thought and effort. But you shouldn't attempt to begin production without it. It's as close as you can come to a guarantee of success.

Who's Going To See It? And Why?

Communications researchers have spent considerable time and money trying to determine when and what type of multi-image presentation is most effective. Unfortunately, their findings haven't been all that conclusive; the studies wind up softened by qualifications or tempered by experience. Older people, state the reports, aren't the best audience for multi-image presentations — unless, of course, they're interested in the subject matter. Children and young adults, the so-called "TV generation," seem to get more out of multi-image presentations — that is, if the length of the presentation doesn't exceed certain limits. "Creative" people grasp a multi-image message more readily than more literal-minded people — on the other hand, they're also more apt to miss the message as they become interested in aesthetics and technique. And to keep audience attention from wandering, researchers advise limiting a presentation to between 15 and 30 minutes — although experience shows that programs of up to three hours can be extremely successful.

So researchers can't supply a handy rule of thumb to use in determining when and how to use multi-image. But that doesn't mean you should simply discard their findings — qualified as they may be. Those findings, coupled with the same kind of common-sense analysis you would use in determining the appropriateness of a motion picture or a simple slide presentation, can help form your decisions.

The information needs of an audience determine not only what a presentation will say but also how it will be said.

What decisions? Go back to the questions we asked earlier. Who is your audience and what are they looking for or expecting from your presentation? Will the use of multi-image enhance your chances of accomplishing your communications objectives — or will it get in the way? Does the occasion justify the use of multi-image — in dollars and cents, time, and effort?

Let's start with the audience. Who are they? What do they want? How might they react to multi-image? You have to answer those questions; you can't rely solely on the findings of researchers. If you're in the business of communicating, **you** have to know something about the people with whom you're trying to communicate.

You've set your communications objectives, so you already know something about your audience. You know if your audience will be voluntary or "captive." You probably know what age groups will make up that audience. And you probably know their positions in life — that they're salespeople, or doctors, or financial analysts, or college students, or just "people off the street." With this sort of information you can begin to determine if message and audience are compatible.

Now let's look at the occasion. It's usually easy to decide if multi-image is suitable for a particular communications occasion. All you have to do is answer one question: Is the multi-image format compatible with the message of the program? In other words, does it make good communications sense to enhance your message with the impact — and costs — of multi-image production?

A few examples will illustrate what we mean. In the first, let's say the occasion is a company's annual meeting, the audience a group of shareholders, and the message an explanation of the company's marketing plans for the coming five years. Is a multi-image presentation appropriate in this situation? Are message and medium compatible? On the basis of the questions we posed, the answer is probably "yes." The impact of multi-image would add an aura of strength and confidence to the presentation, which is the underlying message the company's management would want to get across.

Next let's look at an advertising manager of a toy company who wants to introduce a new line of toys to almost 500 buyers. Are message and multi-image medium compatible for this occasion? Again, the answer would appear to be "yes"—message, occasion, and audience demand multiple projectors, screens, and images. In fact, that's exactly what the people at Mattel, Inc. decided when they used five screens, 17 slide projectors using 2,500 slides, a 35 mm motion picture projector using 6,000 feet (1829 m) of film, strip lights, laser beams, cloud-forming equipment, water fountains, and a quadrophonic sound track to produce "Dimension '78," a two-hour and 45-minute multi-image presentation of toys and marketing plans.

But not all messages demand visual excitement on an elaborate scale. If a communications specialist were asked to prepare a program announcing minor changes in a company's dental health plan to an audience of 30 employees, he or she would have difficulty justifying multi-image as the appropriate vehicle.

So it's hard to formulate a rule of thumb that will apply in all situations. But it is nevertheless accurate to say that multi-image is appropriate for an occasion in which visual excitement and emphasis will support and strengthen a presentation's basic message. But just knowing that multi-image is appropriate for your audience and occasion isn't enough. Before you forge ahead with production, other decisions and plans must be made.

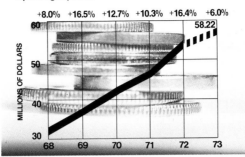

Operating Expenses

The appropriateness of art, slides, screen format and even multi-image programming techniques can be determined by thorough audience analysis. (Photo: Laurence Deutsch Design, Inc.)

21

Presentation Requirements

Once you've determined that multi-image is an appropriate format for your audience and the occasion, you should begin to give some overall definition to your eventual presentation. That means deciding how long the presentation should be. Whether the narration should be "live" or "canned." Whether the presentation should incorporate slides **and** motion pictures. How large an audience will be viewing it (and therefore, how large the smallest image must be if you're considering multiple images). The list of decisions can go on and on, depending on how closely you—as a decision-maker—will also be involved in the production.

Coping with these details and decisions can overwhelm you unless you have some method for sorting them out. You need a means for pinpointing your project goals and giving them a priority. Only by performing this preproduction analysis can you determine if your proposed presentation is feasible in terms of time, talent, and budget.

Although the factors you seek to identify and organize are numerous and complex, the method for completing this analysis can be simple. One approach many decision-makers have found both easy and effective is the **Must/Want Chart.**

The construction of a Must/Want Chart begins with communication and ends with decisions. You can construct a chart alone if you wish, but it's more useful when you involve all the people and groups with an interest in the completed presentation. If, for example, you're producing a presentation for a sales meeting or trade show, you should involve people representing advertising, sales, and exhibits, as well as your audiovisual specialists. If the presentation will be used for training, you should involve a representative from the group requesting the program as well as training and AV specialists. By involving all key people and groups you promote active communications—the process of give and take needed to make decisions and set priorities.

A Must/Want Chart organizes this decision-making effort through the use of a four-step process. You've already taken the first step—**setting your communications objectives.** Listing your objectives in terms of the desired audience reaction helps you sort out the important requirements for your presentation.

Presentation requirements are statements of the elements to be included in the completed presentation or factors to be considered during production. Again, depending upon how deeply you and the other members of the decision-making team will be involved in the actual planning and production of the presentation, the listing of these elements and related factors will be more or less specific.

The major elements to be considered include the site of the presentation; the use of live and/or canned narrations; the use of slides, film sequences, or other visual effects; the size or number of screens to be used. Additional elements to consider might include tne number of projectors to be used, the type of programming equipment to be used, whether the screens will be front or rear projection, etc. Factors to be considered during production include length of presentation, the size of the staff available for production, size of audience, the number of times the presentation will be used, types of images to be shown, secondary uses of the presentation, and budget limitations if they exist.

Step two in the construction of a Must/Want Chart consists of listing all these considerations and any others that may seem relevant. At this stage in the construction of the chart, your guiding rule should be "anything goes." Your goal is to encourage everyone involved in the process to contribute his or her thinking. The last thing you want is for someone to have unexpressed expectations or reservations about the completed presentation. The reason is simple: You can deal with objections or unrealistic requirements much more easily before a production is launched than you can at a later date, when decisions have been transformed into slides, audiotape and programming.

When the listing of requirements is completed, you will probably have anywhere from 10 to 30 statements on your master list. This is too many items to serve as a useful decision-making tool, so further refinement is necessary.

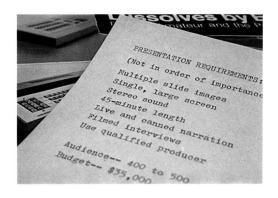

In step three, the presentation requirements are divided into **Musts** and **Wants.** A "must" is an item that will literally make or break your multi-image presentation as originally conceived. If, for example, you decided that your multi-image presentation **must** contain recorded sound effects and narration along with segments of live narration, then any presentation offering less than this should be considered unacceptable. (You can, of course, revise your requirements, as will be explained later.) A "want," on the other hand, is an item you would like to include in your presentation, but would be willing to sacrifice without altering the acceptability of the show. For instance, you might **want** to use film images along with your slide images. But if shortages of time, talent, or money preclude film production, you could still go ahead with your presentation without compromising your primary goals.

This listing of musts and wants is especially helpful in determining budget and time requirements. If no budget exists for a presentation, your job becomes that of putting a price tag next to all the items in your "must" and "want" columns. (How you go about determining these prices is the subject of the next chapter.) The total for the "must" column gives you a minimum figure below which you can't go without compromising your objectives. The figure for the "want" column gives you the cost for adding refinements and polish to the basic presentation.

In much the same way you'll also be able to determine, at least roughly, if you have the time—and the staff—necessary to produce the "musts" and "wants" in your chart. The only difference in the procedure is that you'll be calculating days instead of dollars.

If a budget for the presentation does exist, your job is to determine if it is adequate. It is if it covers, at the very least, the costs for the "musts" on your list. If the budget doesn't cover these costs, then you must decide to abandon your plans for a multi-image presentation or, more likely, you must lower your sights a bit.

Step four of the decision-making process enables you to focus time and money on your more important requirements. If necessary, it also gives you an easy way to modify your original requirements.

In this step you set priorities for all requirements, in both the "must" and "want" columns. The most effective way to set these priorities is to judge your requirements in light of your communications objectives. Knowing what you want to accomplish will help you determine which elements and factors contribute most to reaching that goal.

An example will make this clearer. Let's say your communications objective is to encourage prospects to ask for a demonstration of a new product. Let's also say that two of your "musts" are "to use multiple slide images simultaneously to illustrate product benefits" and "to use film clips to demonstrate the product in operation." How do you set priorities for these two requirements, which when examined out of context seem of equal importance? You go back to the original communications objective that states that you want prospects "to ask for a demonstration." Which of the two "musts" is more likely to accomplish this goal? Although arguments could be made to support either choice, the probabilities are greater that the multiple-slide illustration of benefits would whet a viewer's appetite for a demonstration more than a movie that gives, in effect, the equivalent of a demonstration. So in this example, the multiple slide images would receive a higher priority than the film.

Setting priorities in this way helps you establish production goals and allocate your budget. The numbers next to the requirements tell you where emphasis must be placed and where cuts can be made. If, for example, your budget allows for all of your "musts" but only some of your "wants," you need only cross off the lower priority items on the "want" list and your budget problems are solved.

The setting of priorities also helps you revise requirements when costs promise to surpass budget. When a budget won't allow for all the listed "musts," you have a starting point for rethinking your presentation and modifying your requirements. In deciding what revisions should be made, you start with your lesser priorities and work up the list. When you've cut enough requirements to make the presentation affordable, then you must go back to your communications objective to determine if the revised presentation will meet your goals. If it won't, you may have to revise your objective—or allocate a larger budget to the presentation.

24

As you've probably surmised, the construction of a Must/Want Chart is a dynamic process. Only infrequently will you move from communications objective to requirements to budget without having to pause to reevaluate your thinking. More typically, you'll move up and down this decision-making ladder many times as you refine your goals and your plans. This isn't a sign that your original thinking was faulty; rather it's a sign that the Must/Want Chart is doing its job — it's forcing you to think about a multi-image presentation in all its phases and then to be realistic in your planning for those phases.

The worst decision you can make at this time is to avoid the problems and questions posed during the construction of the chart. To ignore this process — or to attempt to make these decisions "as you go along" — is to invite eventual production headaches. You run the risk of concentrating on "wants" rather than "musts," of spending for lower priority rather than higher priority requirements, of focusing on requirements that don't contribute to your goals.

Of course, it is possible to make all these mistakes and still produce a successful multi-image presentation. But even if you choose to take an improvisational approach to decision-making, you still have a very important initial decision to make — are you willing to trust the success of your presentation to luck?

A Must/Want Chart forces producers to examine — and frequently, to revise — their thinking and planning before actual production begins.

Setting priorities is the critical last step in the development of a Must/Want Chart.

Setting Priorities With A Must/Want Chart

Planning for a multi-image presentation often seems like an exercise in mental juggling. Dozens of factors must be identified, categorized, compared, and ranked, and the very act of performing this analysis sometimes seems to complicate matters rather than unravel them. But a simple form of help is available.

A Must/Want Chart helps you evaluate and give priorities to factors of seemingly equal importance. The method is almost mechanical in application, but the decision-making process it sets in motion is dynamic. It forces you to plan and to make decisions based on this plan; almost always it also forces you to reevaluate your plan on the basis of your decisions. This refining process results in a list of requirements that will serve as firm project guidelines.

The following example shows how a Must/Want Chart is constructed. It is based on the experiences of a manufacturing company that wanted to use a multi-image presentation to kick off a major sales meeting. The audience consisted of the company's sales representatives and dealers. The occasion was a by-invitation-only meeting, held at the opening of an industrial trade show. The message was "our product and marketing plans for next year mean increased business for those who actively participate in our advertising and sales programs."

Here's how the manufacturing company put together its Must/Want Chart:

1. Set a communications objective.

Knowing that a communications objective must be based on a change in human behavior or thinking, the company, after several attempts, settled on this statement: "To enduce attendees to visit our corporate exhibit, where they will ask for additional details about our upcoming marketing programs."

2. List all presentation requirements.

The initial meeting to discuss this project consisted of representatives from sales, advertising and exhibits, plus specialists from the audiovisual communications department. In what amounted to a brainstorming session, the participants developed the following list of requirements:

Presentation Requirements

- multiple slide images
- single, large screen
- audience—400 to 500
- stereo sound
- 45-minute maximum length
- live and canned narration
- secondary uses
- filmed interviews
- budget—$38,000
- use qualified producer

3. Separate requirements into "musts" and "wants."

It's at this point that decisions were made. The group had to decide what was necessary for the success of the presentation and what was merely desirable. In light of its objective, the group settled on this breakdown of "musts" and "wants":

Musts	Wants
• multiple slide images	• stereo sound
• 45-minute maximum length	• filmed interviews
• canned narration	• single, large screen
• use qualified producer	• live narrator
• images large enough for audience of 400-500	• secondary uses
• budget—$38,000 max.	

4. Set priorities for all "musts" and "wants."

To bring its requirements into sharpest focus, the group next set priorities for both lists. Here again, they used their communications objective as the touchstone against which to measure the ultimate contribution of each requirement.

Musts	Wants
1. images large enough for audience of 400-500	1. filmed interviews
2. multiple slide images	2. single, large screen
3. budget—$38,000	3. live narrator
4. 45-minute maximum length	4. stereo sound
5. use qualified producer	5. secondary uses
6. canned narration	

At this point, having ranked its requirements, the group was prepared to evaluate the lists in light of time and budgetary limitations.

This example will not carry their analysis and decisions any further; that requires material not yet covered. Nevertheless, on the basis of what the chart alone tells us, it's obvious that a big-screen, multiple-image presentation is the group's overriding requirement. All other "musts" are secondary to this. Even an increase in the budget could be considered if it were necessary to provide for these primary "musts." It's equally obvious that considerations of secondary uses for the presentation need not enter seriously into the planning for the project. These and other factors become apparent—and decisions can be made more readily—because the Must/Want Chart forces you to make intangibles tangible. It gives objectivity to what is otherwise a very subjective process.

Evaluating Your Resources— Do You Have The Time?

There's a famous cartoon—variations of which hang in art studios, ad agencies, and AV houses across the country—that depicts a man behind a desk or drawing board with a look of disbelief on his face. The caption usually says something like "You want it **when**?" It's a universal reaction to what always seems like a request for too much too soon. And under few circumstances is that pained question apt to be asked with more disbelief than when it relates to multi-image production.

Time. It's an element that can't be altered. Just as the scope of a multi-image presentation must often be curtailed to stay within a budget, the sophistication of a production must often suffer for lack of time. And more often than not, that lack of time was brought about by unrealistic scheduling or, what's worse, a total lack of scheduling.

If you as a decision-maker, or as a member of a decision-making team, are going to spend the time setting objectives and structuring requirements for a proposed multi-image presentation, it would be foolish to ignore the obvious question: Do you have enough time to complete the production before the deadline?

You can—and should—get a rough answer to this question shortly after you've prepared your Must/Want Chart. The chart will identify the basic elements of your presentation, and from that information you'll be able to project a "critical path" to completion of production. In Section Two—Planning—you'll find more detailed explanations regarding the actual scheduling of time for various functions involved in multi-image production. But for now—while you're still at the stage where change is less expensive and less traumatic—it's important to estimate just how much time will be required to produce the show you're planning.

At this juncture, it would be safe to break the production schedule into five major sections:

- Research, treatment, and scripting
- Design, art, photography, and/or cinematography
- Sound track and visual assembly
- Programming
- Rehearsal

There are, of course, no hard and fast rules that dictate this specific sequence of events, or even that each section must be completed before another is begun. But in general, because they involve one or more distinct skills, each section is usually dependent upon the preceding ones for direction. By analyzing the time requirements for each of these production specialties, you'll be able to develop a production schedule much as you would develop a cost estimate.

The result of this analysis won't assure you a clockwork production, nor will it guarantee that you'll avoid a last-minute panic situation. However, it will at least provide the assurance that your efforts aren't being directed towards an impossible goal.

Matters Of Money

When multi-image producers find themselves explaining their business to people with little or no experience in the field, they usually brace themselves for the inevitable question. The producers know the question's coming, and they know it's probably the most important question in the other person's mind. They also know they don't have an answer that will completely satisfy the questioner.

And that question is: How much does it cost to produce a multi-image presentation?

This seems like a simple question, but it's not. And if multi-image producers seem evasive when answering the question, it's because they have no other choice.

The truth is no one **can** answer that question. It's like asking how much does it cost to build a house? It can cost $10,000, $100,000 or $1,000,000, depending on how big and how luxurious a house you want built. The same is true with a multi-image presentation. Production costs depend on the length of the show, the complexity of the show, the equipment needed to produce and present the show, the site and layout of the theatre where the presentation will be shown, as well as dozens of other factors.

What all this means is that you can't talk about costs until you can talk specifics.

It's usually at this point that the novice says to the professional, "Yes, I understand all that. But isn't there some rule of thumb you can give me to help me estimate my costs?"

Again the answer is "no." People who make their living by producing multi-image presentations have given this question a great deal of thought. They would like to be able to tell prospects that a production will cost so many dollars per minute or per projector or per screen or per image. It would make their bidding procedures a lot easier if some reliable guidelines existed to help in estimating. But experience has taught them these "cost guides" don't exist. And even if some rough standards did exist, only a foolhardy producer would stake his reputation and profits on their accuracy.

The problem with "standards" or "guidelines" or "rules of thumb" is that they're based on averages, and there's no such thing as an "average" multi-image presentation. The options available to a producer — in number of screens, number of simultaneous images, number of projectors, number of sound tracks, type of programming equipment, or sophistication of art or photographic techniques — introduce too many variables to be served by a constant "average." So, the experiences of independent multi-image producers reveal this fundamental lesson: To estimate the cost of a multi-image presentation, you must begin with specifications for a particular multi-image presentation. There's no other "easy" way.

You, as a decision-maker, won't have a problem developing the specifications from which to calculate a budget. In fact, these specifications already exist, in broad terms, in the entries on your Must/Want Chart. If prepared correctly, that chart should tell you the length of your presentation, as well as the general style of the presentation. Your next step is to assemble, on paper, the staff and equipment you'll need to bring these requirements into existence. Once that's done, you have to put a price tag on these resources.

Before you tackle this assignment, however, consider this bit of advice: Unless you're experienced in audiovisual production, don't attempt this planning and budgeting alone. You're going to need the help of an expert, someone who's familiar with the skills, equipment, and costs involved in audiovisual production. This person might be someone on your organization's staff, such as an audiovisual specialist. Or it might be someone you'll have to hire as a consultant—an independent producer or a production specialist from your advertising or public relations firm.

Once you've found this "technical adviser," the actual process of estimating production costs can be simplified by examining the project in terms of the resources needed.

Skills

The easiest resource to begin with is skills—or to look at it from a more practical viewpoint, the people you'll need to handle various production assignments. Under this category you may be budgeting for any or all of the following production skills:

- scriptwriters
- photographers
- cinematographers
- graphic artists
- animators
- sound recording specialists
- sound engineers
- lighting specialists
- stage directors
- programming specialists
- decorators
- motion picture editors
- narrators
- stage actors
- photographic lab personnel
- projectionists

Which of these skills will you need? Turn to your Must/Want Chart for the answer. Is your presentation going to have narration? If so, you're going to need a scriptwriter. Almost certainly you'll need a still photographer, maybe even more than one. Again your presentation requirements should indicate, at least roughly, the amount of photography involved in the project. Using motion picture sequences? Then you're going to need the services of a film crew—minimally, a cinematographer and a motion picture editor. If the sequences will include on-location sound, you're also going to need someone to record, edit, and mix your final sound track.

As you've probably surmised, common sense will identify the specific skills you'll need to pull your presentation together. But don't stop your planning at the production stage. If you do, you'll be committing one of the most frequent mistakes that inexperienced multi-image producers make—failing to budget for the skills needed to set up and put on a presentation. Naturally, you're going to need someone—maybe more than one person, depending on the complexity of the final production—to run the presentation equipment. Keep in mind that in some locations you may be required by local unions to use their members (for example, projectionists) to run the presentation. You're also going to need people (who may also be union affiliated) to prepare your presentation site. Depending on where your presentation will be shown, these skills could include:

- carpenters (to construct a projection booth and other displays or exhibits)
- electricians (to provide electrical hookups for all your projection and programming equipment)
- interior decorators and drapery hangers (to prepare the presentation site)
- specialists in audio systems and screens

What will these skills and services cost? This is a little harder to determine; but with the help of your production specialist, you should be able to develop realistic budget estimates. The key to developing these cost figures is to divide your skill needs into those you can satisfy with your present staff and those you'll have to provide for from outside sources.

Budgeting costs for in-house specialists is relatively simple: Your production consultant should be able to estimate the probable amount of time needed to complete the various production assignments. With these figures and salary costs for each of the in-house specialists, you can easily calculate that portion of your production costs.

Developing cost estimates for outside services is somewhat harder. Unless you or your production consultant have recently dealt with a supplier on a similar project and can thus use actual cost figures as the basis for budgeting, you're going to have

to ask several suppliers to give you estimates. In seeking estimates at this stage of the project, your best approach is to tell the supplier what sort of skills you'll need, for how long, and for what types of work. If, for example, your presentation calls for motion picture sequences (with synchronous sound) of a parade or sporting event, give all this information to a local producer and ask for an estimate.

In doing this, explain to the producer that you're just pulling together figures for a budget and that, if you were to begin actual production, the producer would be asked to bid again on the basis of a more detailed description of the project. This is only fair to you **and** the producer; neither party would want to be held to cost figures developed during an initial budgeting phase.

Only one other point needs to be mentioned concerning skills. The budgeting procedure just described assumes that someone from your organization will be handling overall production responsibilities. This doesn't mean you'll necessarily have all the skills you need under one roof. More than likely you won't; you'll be hiring specialists such as a cinematographer and a sound engineer. But they will be under the direction of an in-house producer or coordinator, who will make assignments, set deadlines, and evaluate results.

The other option available to you is to hire an independent producer, who will assemble the talent needed to complete your production. In this case, **you** don't have to estimate costs for each skill. All you need is a total production estimate from an experienced producer.

Equipment

In much the same way you identified the skills you'll need to complete your production, you also have to go through your Must/Want Chart to identify the equipment you'll require. Following is a list of the most common equipment needed to produce and present a multi-image program.

- 35 mm still camera with lenses, tripod, and strobe unit
- flood lights
- copystand
- slide illuminator/organizer
- slide trays
- dissolve control units
- slide projectors with spare lamps and projection lenses
- projection stands
- motion picture camera and lenses
- motion picture sound recording gear
- motion picture editing equipment
- motion picture projector
- ac power relays
- tape recorder
- amplifier
- loudspeakers
- screen(s)
- programmer
- audio mixing equipment

The equipment needed for a presentation can be determined using a Must/Want Chart.

Of course, unless yours is a new organization, you probably have some of this equipment. Many pieces of equipment, such as 35 mm still cameras or reel-to-reel tape recorders, are a natural extension of the people who use them. So, if you have a photographer and a media production specialist on your staff, you probably have the equipment they normally use. Other pieces of equipment, such as 2 x 2-inch slide projectors and 16 mm motion picture projectors, have found such common application in business and education that you probably also have these and similar products on hand.

Of course, the equipment you don't have you'll have to obtain, and that poses a basic question: Should you buy or should you rent? Answering this question will require you to step away from your current production—at least for the time being. You're going to need a broader perspective to answer this buy-or-rent question, one that includes your organization's long-term communications requirements.

The reason for this long-term evaluation is simple: On first consideration, as you face the pressures of a current production, the option of renting equipment seems the more economical answer. A piece of equipment as common as a 16 mm motion picture projector can cost from $700 to $4,000. More sophisticated equipment, such as programmers, can cost from $1,200 to $12,000. You can rent equipment like this, on the other hand, for a fraction of the cost. So, when faced with such divergent choices, your first reaction may be to rent. That decision keeps the budget for a current production within reasonable limits.

And the decision is a good one—if your communications plans are limited to a one-time-only production. But if audiovisual communications—especially multi-image communications—seems likely to form a growing part of your total communications activity, then the decision to rent could, over the long run, wind up costing you more. Also, even though rental charges for a particular piece of equipment may seem favorable on the surface, you should recognize the fact that any delays in your production schedule may increase the total rental costs substantially.

So the decision to buy or rent is a difficult one, and no guidelines—beyond your organization's communications requirements—exist to make the job any easier. Foresight, a calculator, and some common sense are the tools you'll have to use to reach an answer. But while no guidelines exist, the following suggestions may help you clarify your thinking.

Buy equipment you'll use on a regular basis. Most organizations using audiovisual communications require certain basic equipment. This equipment would include:

- 35 mm still camera with a minimum of four lenses—normal, macro, wide-angle, and telephoto; tripod; strobe unit
- 2 x 2-inch slide projectors with zoom lenses
- a basic dissolve unit (for use with simple two-projector slide presentations)
- a slide illuminator/organizer
- a reel-to-reel tape recorder with editing capability
- a cassette recorder with built-in slide synchronizing capability for simple slide presentations
- a basic copy stand
- peripheral equipment such as extension cords and equipment carrying cases or shipping cases

This is the type of equipment your organization will use so frequently that it would prove uneconomical to rent it every time a need arises. The cost of this equipment is also reasonable enough that it can usually be purchased without major capital expenditures.

Rent equipment you'll use only occasionally. Your organization's requirements—present and future—will identify the items in this category. For most organizations, renting makes greater economic sense with such equipment as:

- 16 mm motion picture projectors
- 16 mm motion picture editing equipment
- double-system sound recording equipment
- large, theatre-size screens
- special projection lenses for 2 x 2-inch slide and motion picture projectors
- projection stands for slide projectors
- speakers and amplifiers for theatre sound systems
- four-channel tape deck for stereo sound

These pieces of equipment are needed for the production and presentation of more sophisticated audiovisual and multi-image communications programs. Of course, if your organization produces such shows on a regular basis, many of these units could be bought rather than rented.

Purchase certain services rather than rent equipment. Some equipment used in multi-image production is so specialized that even renting it makes little sense. One such piece of equipment is an animation copy stand. This unit is used to film or photograph artwork or photographic prints or slides in a controlled, precise way, and it takes an experienced operator to use the unit correctly. If you find you need slides or film sequences copied with such a unit, your best approach is to contract for the services of both equipment and operator.

The toughest decision—your programming equipment. Should you buy or rent multi-image programming equipment? That's the toughest question you'll have to answer, whether you're a first-time producer or someone with several shows behind you. In fact it's probably more difficult if you have experience, because only then do you truly appreciate the options—and obstacles—lying before you.

If this is your first venture into multi-image production, you're probably better off to rent programming equipment. You might even consider contracting for the services of an experienced programmer, who'll also supply the necessary equipment.

The complexity and diversity of programming equipment makes the decision to rent advisable for the novice producer. The problem is, the novice doesn't know what he or she doesn't know. In fact, the novice is often surprised to find out the term "programming equipment" doesn't have much specific meaning. To say you need "programming equipment" is the equivalent of saying you need a "vehicle" when what you really want is a Chevrolet Camaro, with AM/FM stereo radio, air conditioning, automatic transmission, custom wheels, tachometer, and sports mirrors.

"Programming equipment" is a generic term, an umbrella used to cover several basic types of equipment, manufactured by dozens of companies, and containing a wide variety of features. So, choosing a programmer involves more than merely selecting Model A over Model B. You must decide if a paper-tape programmer or a multi-frequency tone programmer or a computerized electronic programmer is best for your needs. And only experience will enable you to make that choice—as well as the choice of features offered by different units within each category. So buying a programmer when you first begin programming—even if your organization's requirements can support the purchase—can be a costly mistake; after two or three productions, you're likely to find yourself in one of two situations: Either you've purchased an expensive programmer far too sophisticated for your needs, or you've purchased a programmer too limited in its capabilities. The result in either case is discouraging.

Once you have enough experience to know what you need from a programmer, then your problem becomes one of justifying its cost in relation to its use. How much you're willing to pay should be determined by the number and sophistication of the shows you'll produce every year. The greater your requirements, the more you should be willing to spend. Obviously, if you produce one or two relatively simple shows a year, you don't need—and can't justify—a $12,000 computerized programmer that controls the functions of 50 projectors. A simple three- or four-channel programmer, with a price tag of less than $1,200, may be all you need. On the other hand, if sophisticated multi-image productions will form a major part of your organization's communications activity, then several computerized programmers may be easily justified.

Of course, justifying the cost of programming equipment usually comes down to a subjective decision you must make. It's not always easy, especially if the purchase involves more sophisticated equipment. Advances in microprocessing have revolutionized programming technology. In fact, the marketplace has become a competition of "technological leapfrog," with manufacturers introducing programmers that are smaller, more versatile, and more sophisticated than those introduced only a year earlier. Justifying this sort of expensive and rapidly outdated equipment is difficult unless you use it on a regular basis—as the manufacturers themselves will point out. They would rather have you buy a less sophisticated—and less expensive—piece of equipment that meets your needs than a top-of-the-line unit you'll never put to full use. So justifying the cost of a programmer usually comes down to one question: How long will a particular programmer's capabilities serve my organization's needs? Divide the purchase price of the programmer by this estimate of useful life. If the resulting annual "cost" is one you can justify in terms of use, then go ahead and make the purchase. If you can't justify the purchase, look for a less sophisticated model.

Developing cost estimates for all the equipment you'll need will take some time, but that doesn't mean it's a difficult task. Ask two or three audiovisual dealers in your community to give you quotes for the equipment you have to buy. Then, in budgeting, use the prices from the dealer you're most likely to do business with.

For the equipment you'll rent, the procedure is similar—ask for prices from companies that rent audiovisual equipment and use the prices from the most likely supplier.

Supplies and Materials

The next items you must budget for are the consumable supplies, materials, or services needed during production. Common items in this category include:

- film (for motion picture and still photography)
- film processing
- audiotape
- slide mounts
- slide masks
- write-on slides

The quantity of each item you'll need depends on the length and scope of your proposed presentation. Naturally, the greater the length and complexity of your show, the greater your material costs are going to be. Once familiar with your overall goals, your production specialist will be able to develop a cost estimate for this category.

The degree of sophistication necessary in a programmer depends to a large extent on the number and sophistication of the presentations planned for production.

Facilities

One area of potential costs almost always overlooked by novice multi-image producers is facility or "space" costs. It's overlooked because the novice assumes that, since everyone already has working room, there's no need to supply additional space.

Once you've produced a multi-image show, however, you'll recognize the fallacy in this assumption. A multi-image production is more than just an inflated slide show; it's a medium with special requirements, and one of them happens to be a large production work area.

The production of a multi-image presentation usually requires a large work area, a cost often overlooked by novice multi-image producers.

Space is needed for slide editing and mounting. How much space you'll need depends, of course, on your production requirements. If yours is a limited production, you'll need limited space. But if you're planning for a show using nine or more projectors, you had better plan to find additional room. How much room? Well, for the Mattel, Inc. show mentioned on page 21, the producers shot 18,000 slides, then edited this number down to the 2,500 used in the presentation. That kind of effort requires a lot of room. Your needs probably won't be as extensive. Nevertheless, your photographer or audiovisual specialist will need room to review, sort, edit, and mount the hundreds of slides required for even a modest multi-image presentation.

Space also must be provided to project the show **as it's being assembled.** The producer will want to see sequences of the presentation as they come together — on a screen that proportionally duplicates the dimensions and configurations of the screen or screens to be used at the public performances. Providing this space won't create many problems if you're planning for a simple two- or four-projector show. But if you're producing a show using a large number of projectors, you need to provide special space for this projection requirement.

This additional space is needed for two reasons. First, projecting images from a large number of projectors onto a single screen in a normal-size room is going to cause a problem called "keystoning." That means the images from the projectors on the outside of the projection setup will tend to strike the screen at such a nonperpendicular angle that one dimension will be longer than the other. For a multi-image producer concerned with the alignment and registration of multiple images, keystoning is an undesirable condition.

The second reason a special projection area should be provided is to prevent anyone from moving or tampering with the projection equipment. As just mentioned, one of a producer's key requirements is proper alignment and registration of multiple images. To achieve this registration, a producer may spend hours setting up and positioning projectors. And, of course, the more projectors involved in the production, the more time that must be spent on this task. So you can see the problems that could arise if this projection room were "open to the public." Someone needing a projector for a few hours might be inclined to borrow one from the room. Or someone holding a meeting might want additional space for chairs or charts and simply move some of the projectors "out of the way" for a few hours. Such innocent acts could mean many additional hours of work for the producer, spent realigning and reregistering the projectors.

If your operation has a large, unused area—or if you plan to have the presentation produced on the outside—you don't have to figure any additional costs for space. But if you need space for an in-house production/projection room, you may have to budget for costs involved. These costs might include interorganizational charges for renting the space, and costs incurred if a maintenance staff has to provide additional electric power or equipment to make the room usable.

Your organization's facility management staff will be able to supply an estimate of these costs.

Shipping

The final category of costs you must budget for includes those incurred in shipping your show to the presentation site. Naturally, if the show is to be presented in your offices, few—if any—costs will be involved. But if the show is to be presented at a site outside your facility, you must budget for the costs involved. These costs could include:

- the construction or purchase of packing cases
- transportation costs
- insurance costs

Costs for shipping mount the further, more often, and faster your presentation must be moved. If the show is to be presented locally, your organization can probably handle all transportation activities. Your costs will result in an interorganization charge.

If the show is to be presented out of town, however, you'll probably have to contract for the services of a shipping or freight company. Here again you should be able to turn to someone on your organization's staff for help. Most organizations, whether they're involved in business, education, or government, have a specialist in shipping or distribution. Once this person knows your requirements, he or she will be able to estimate your costs.

As a bonus, a distribution specialist will also be able to organize the logistics and scheduling involved in transporting your presentation.

Shipping is another cost often overlooked.

The Bottom Line

Add up all the cost estimates you've gathered. Then add 10 percent of that total to the estimate for "contingencies." That additional money should cover the one or two small items you forgot to budget for—and they always pop up! It will also cover the unanticipated problems you'll encounter along the way—they too are inevitable! This final figure—the bottom line—is what your planned presentation is going to cost.

Now, if you're like most decision-makers, you're going to look at that figure and ask a very natural and logical question: Is there any way I can save money?

You're going to ask that question either because you already have a budget you're working against—and have already exceeded on paper—or you'll ask it because you feel first budget estimates are typically too high and that there must be a way to cut back the costs.

Be prepared for unexpected production costs by budgeting a percentage of the total budget for contingencies.

And you're probably right on both counts. The budget more than likely is too high, and there are ways of cutting the estimated costs. To make these cuts—and to get the budget to the point where you feel comfortable with it—you must go back to your **Must/Want Chart** with one thought in mind: Is there any way to simplify these requirements without jeopardizing the communications objective?

In asking this question you'll be forcing yourself—and the others involved in the planning process—to examine the various production options available. And there are a number of them. As mentioned earlier in this chapter, a multi-image show is the product of dozens of variables, so every time you simplify those variables you're probably simplifying your show.

You can simplify your presentation—and reduce your budget—by cutting back on the number of media, projectors, screens, and simultaneous images in your original requirements. You can reduce the length of the presentation. And you can look at your equipment list with the idea of determining the minimum capability needed to achieve the results you want. This type of can-we-do-it-with-less thinking applies to items ranging from the lenses for a still camera to the sophistication of programming equipment to the durability of the shipping cases you construct or buy. The reason for this examination is to remind yourself that you don't necessarily need the biggest or the best or the latest. You need equipment that will do the job **you** need done, at the least possible cost. It's not effectiveness, but **cost effectiveness,** that should guide your thinking.

Where you simplify and how much you simplify will be determined by your presentation requirements. And in 95 cases out of 100, you should be able to simplify. That's because the bottom line on budgeting for multi-image production is this: It is imagination more than a room full of expensive equipment that makes multi-image effective.

How To Choose
An Independent Producer —
If You Need One

For many decision-makers the task of choosing an independent multi-image producer seems mysterious and often forbidding. They become inhibited by doubts and questions: How do you go about it? Where do you start? What questions do you ask? What factors do you use to narrow your choice to one? The questions are logical and natural. The problem is many decision-makers don't think they know the answers to those questions.

But for the most part they're wrong. That's because hiring an independent producer is very much like hiring a public relations consulting firm or an accounting firm or, in some cases, a new employee. It's like hiring or retaining an outside consulting firm if you're going to give a producer responsibility for the complete production. It's probably more like hiring a new employee if the outside producer is going to work with members of your internal staff during production. In either case, for the period the producer will be working for the decision-maker, he or she is the equivalent of an employee. During the production process, the multi-image producer is a very real part of your staff, trying to meet your communications goals, solving the communications problems you need solved, and producing the show you want produced.

So there are no special interviewing techniques to use, no magic formulas to follow. However, if you are looking for a producer who will provide creative direction and technical know-how — but will use production specialists from your staff — there are a few questions you should ask yourself before you begin the search and selection process: Is this talent going to be a frequent requirement? Is it necessary to hire someone from the outside? Or can someone on staff take on these responsibilities? Can someone be promoted to the responsibilities?

If you find someone on staff with the time and potential talent to handle the job, **and if multi-image production is not a pressing need but only a goal,** encourage this person to develop his or her production talents. The individual will certainly need time to gain experience with multi-image production; he or she may even want to take courses to speed up the learning process. But if time allows for this developmental process, then let it to go forward. Having a competent multi-image producer on your staff offers definite advantages. The staff producer will enable you to offer presentations that you wouldn't go outside to have produced. And when the time comes that you want to use an outside producer, an in-house specialist can help you evaluate and choose one and later help supervise his or her activities.

When you decide to go outside for a producer, your first step is to prepare yourself the way you would for a job applicant interview. This means you have to identify the skills and talents you need, outline the scope of the assignment to be completed, and determine the lines of communication and control the producer must work within.

Identifying the skills and talents you need is a safeguard against overhiring. Multi-image producers don't form a homogeneous group, with similar skills and experiences. They all have their own strengths, styles and specialties. Some specialize in theatre-type presentations; others in presentations used in displays and exhibits. Some specialize in sales meetings; others in presentations for specific industries; while others pride themselves in being able to produce for any company, in any industry, for any occasion. Some producers retain their own production specialists, others hire free-lancers to handle specific assignments.

Look for outside producers whose experience and capabilities fulfill presentation requirements.

39

To pick a producer from this diversified group isn't complicated—if you rely on common sense. Take the approach of a wise shopper: Buy only what you need. Just as you wouldn't hire an automotive engineer to change the oil in your car, you shouldn't hire a multi-image producer whose talents and specialties lie outside your requirements.

How do you identify your specific requirements? Go back to the requirements in your **Must/Want Chart.** There you'll find **indications** of the talents and skills needed to complete your production.

Look at what you want to accomplish; then decide what skills you'll need to meet your goals. If, for example, one of your "must" requirements is the production of several shows for use in a large exhibit area, you can conclude that one of the factors you'll be looking for is experience with exhibit area presentations.

Outlining the scope of the production assignment requires you to think through the question of just **what** you'll want the producer to do. Will the producer be responsible for all aspects of the production, from concept to presentation? If not, what portions of the production will the producer handle? And who will handle the other portions? How will the producer work with the others involved? Will he or she work directly with them? Or through you? Or through a coordinator?

The answers to these questions form the equivalent of a job description. They tell you what skills and experience must be brought to an assignment. They also indicate the degree of involvement the producer will have, and his or her relationship to others involved in the production.

You should discuss these working arrangements with producers you're interested in hiring. Depending on how you've outlined the scope of your production, some producers may politely refuse to bid on your project. These producers don't like to get involved in a project unless they have total control over the creative process. They don't like using a client organization's writers, photographers, artists, programmers, or sound engineers. It's not that they don't respect these people. It's just that a multi-image presentation is more than the sum of its parts. It's the product of a creative process, one that many producers feel functions best when writer, photographer, director, and producer work closely with each other, from beginning to end, sharing ideas, problems, solutions and responsibility for results.

Other producers may not object to working with people from your organization. But they'll want to know **exactly** how this working relationship will be structured. And a general description of how things "should fall together" won't satisfy them. They know from experience that the biggest problems they encounter usually aren't linked to the production process. It's unexpressed expectations—those the client has of them or those they have of the client—that usually lead to problems.

But these problems are easily avoided. With a specific description of the proposed assignment in hand, you can tell a producer exactly what you expect and what he or she can expect of you and your organization.

The third decision you must make before beginning the search for a producer is to determine how responsibility and control for the project will be handled. This decision, of course, depends in large part on how you've outlined the scope of the project. If you're hiring a producer to handle all aspects of production, responsibility and control will flow from you to the producer. If others are involved in the project, the lines of authority and control become a little more complicated. You may wind up with the producer and several in-house specialists reporting to you, in which case you become a conduit for information flowing between both groups. Or you might name an in-house coordinator, responsible for the work of your staff specialists and for coordination with the outside producer.

Regardless of how you establish these lines of control, they should be set up **before** you hire a producer. He or she should know how the scripts and recommendations are going to be channeled and how your organization is going to relay information, decisions, and approvals to him or her. For one thing, your approach to organizing this control may well influence the producer's decision to accept or refuse the assignment. But more importantly, waiting until problems arise to establish control can waste valuable time, budget, and energy. Squandering these resources to solve unnecessary problems will result in frustration and discouragement for you and the producer. But in the end it's the presentation that suffers most.

Once you know what you want from a producer and how you're planning to work with him or her, your next task is to choose the one from the many. Here again the approach you'll follow is similar to that used when evaluating a potential employee. That means you'll be hiring on the basis of experience.

How do you judge a producer's experience? Perhaps the most efficient and effective way is to seek out producers of presentations you've seen and found impressive. If you've been to a multi-image show that made you think, "That's just what I want to do for our next sales meeting," find the producer of that show and invite him or her to your office.

Another means of finding a producer who meets your requirements is to ask for recommendations from colleagues and people in other organizations who have used multi-image communications effectively. They'll tell you who produced their presentations and to what degree they found the producer—and their **own** experience—successful. Out of these conversations will come names you can add to your list of candidates.

A third way to identify possible producers is to read reports on multi-image presentations in such magazines as **Audio-Visual Communications, Photomethods** and **Multi-Images: Journal of the Association for Multi-Image.** These reviews usually describe the creative approaches taken by producers and the techniques and equipment used during production and presentation. If a particular approach or technique seems to point in the direction you're thinking, the producer is probably worth contacting.

Another approach is to look under "Audio-Visual Production Services" in the Yellow Pages of your telephone book. There among the lab services and sound studios you'll find listings for creative producers, multi-media producers, multi-image producers, and "all-media" producers. Within these descriptions, you should be able to find a producer with the type of experience you're looking for.

These four sources will give you a preliminary list of producers to work from. Your next step is to invite the producers to show you their work. These screenings will allow you to judge their work from two viewpoints. First, you can judge the aesthetic value of their presentations. That means you'll be looking at the quality of the writing, photography, sound reproduction, and programming, to judge how each element works, both individually and as a component of the overall production.

Second, you can judge the communications value of the presentations. For each presentation you see, ask what communications objective it was meant to achieve. Then judge for yourself if that goal was met. By doing this, you'll be able to distinguish between presentations that "look good" and those that produce results as well.

Once you've talked with several producers and evaluated their work, you must select the one whose style, methods, and experience are best suited to your particular needs. Since there are no objective standards to use when making this choice, you must decide based on subjective feelings. If a particular producer seems right for you, that's the one you should hire.

In cases where you can't decide among two or three producers, you can ask each of them to submit a proposal for the project you're working on. These proposals should cover the creative approach each would take, a general description of how this approach would be developed, and an estimate of the projected costs. On the basis of these proposals, you should be able to select the producer who's right for your assignment.

When you've found and hired your producer, put him or her to work—as soon as possible. What you want to avoid are situations in which the producer begins work after critical decisions bearing on the presentation have been made. These decisions may limit his or her creative options or in other ways restrict the production process. For example, if you decide, on your own, that a particular portion of your presentation must be projected onto a ceiling-mounted screen, you may be placing an unnecessary production burden on the producer, as well as barring the producer from developing a more effective treatment of the material.

You also want to avoid situations where the producer is called in after problems have developed with a presentation. To solve problems, you need a mechanic, someone who can twist, bend, alter, or in some other way modify a situation that already exists. A good producer can do this, of course, but would rather begin at the beginning.

Multi-image production is more art than technology, so the artist you hire should be allowed to work without undue interference or unnecessary problems. The only way this can be assured is to have the producer ready to begin when the project is ready to begin. Otherwise you're paying for talent and experience you're not using.

What Next?

You're ready to go. You have your communications objective, your presentation requirements, a budget, and a schedule. You've selected a producer, either from your organization or from the outside. The next step is to begin.

There 's an art to beginning a multi-image production. It combines equal amounts of motivation, determination, and decisiveness. You want to "kick off" production, in the true sense of what that term implies — supplying sufficient force to launch the project in the desired direction.

You can apply this force to a project at an initial production meeting in which you establish the ground rules and lay out the schedule to be followed during production. Everyone involved in production must know what he or she is expected to do— when, how, why, and with whom. Lines of communication must be explained. Sources of decisions and approvals must be determined and agreed upon.

What all this amounts to is tying up any loose ends that might knot up production at a later date. Once this is done, the preliminaries are over. The initial decision has been made and implemented. The time has come to produce your multi-image presentation.

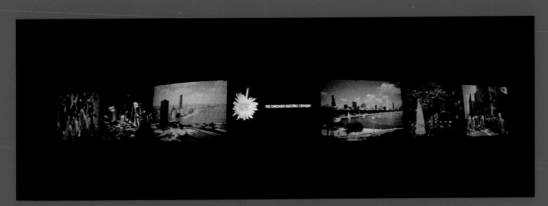

Paul Condylis, producer of the seven-screen production **Chicago Odyssey**.

Paul Condylis:
"Go Ahead And Try It"

Paul Condylis has encouraging advice for newcomers to multi-image production:

You don't have to be a seasoned professional to be successful. All you need are good taste, common sense and a feel for the process of effective communication. If you have these qualities, he says, then take your idea and try it.

That's experience speaking—in a strong, confident, authoritative voice. With a career that stretches back to the late 1940s, Paul Condylis easily qualifies as a consummate multi-media professional. His best known multi-image production is **The Chicago Odyssey**, a 52-minute presentation (Condylis calls it "positive entertainment") examining Chicago's colorful and proud history. The presentation is shown in a 300-seat theatre Condylis had built to house Odyssey's 70-foot screen, a wraparound sound system consisting of 10 speakers, and a 50-foot long projection booth holding 27 slide projectors, three 16 mm movie projectors, a four-track tape recorder, nine United Audio Visual Screenmaster Mark V dissolve units and a United Audio Visual Cuemaster Mark 60 programmer.

More than 2,000 slides, 2,500 feet of motion picture film and a sophisticated light display that recreates the Great Chicago Fire of 1871 are used in the production.

A presentation of the magnitude of Odyssey is beyond the realm of the novice producer, but the thinking that goes into such a production is not.

"As much as we like to have formulas and to be scientific," he says, "there are no formulas for deciding on the style and complexity of a multi-image presentation. These decisions are made on the basis of what you're trying to do."

Condylis lists several factors that must be considered when deciding on the nature and scope of a multi-image presentation.

First—and "primary," says Condylis—is budget. This factor, more than any other, sets practical limits on what you can hope to achieve.

Second is your communications objective and the subject matter to be developed. Do they lend themselves to effective development and presentation through multi-image? If so, what style and visual format are best suited to your message? The answers to these questions will represent subjective judgments on your part, but, says Condylis, your common sense is as good a guide as any.

Third is the site of your proposed presentation. The size of your audience and the setting for the presentation will set limits on what you can reasonably attempt with multi-image.

The fourth and most subjective factor is taste. What can you do tastefully within your presentation requirements? "There's no right answer or best answer to that question," says Condylis. "Asking it is like having asked Picasso if he would paint his next picture on a canvas, a vase or the back of his hand. He would have answered, 'God tells me.' That's the way artistic taste works. It comes from inside, not from formulas."

Once you've examined these factors and made your decision to produce a multi-image presentation, then, says Condylis, let your imagination and common sense come into play.

"Don't be intimidated by the seeming complexity of multi-image," he says. "If you feel in your gut that a multi-image show ought to be a certain way, then go ahead and try it."

"And if you're an executive and not a producer, don't let that stop you," he adds. "Sit down with your creative people and say, 'How can we do it this way? What would work best?' Just don't worry. The right answers will spill out."

Paul Condylis knows this to be true. For him, it's the voice of experience speaking.

43

SECTION II
Planning

This section examines the decisions and plans you must make before launching actual production of a multi-image presentation. This examination will help you refine the considerations expressed in the Initial Decision. You'll learn how to plan for a wide range of subjects, everything from media and visual format to seating, power supplies, and sound. Then you'll learn how to organize your plans, schedule production, and finally how to communicate your decisions through a project proposal and a treatment.

THE BASIC PLANNING DECISIONS — THE BUILDING BLOCKS

Put yourself in this picture:

You're in a meeting at which your department has decided to produce a multi-image presentation. The site and occasion have been set. The objective has been determined. The budget and schedule have been agreed upon. Now it's up to you to begin work on the production. Your first task, you tell the group, will be to develop a plan for production.

Having announced your intentions, what would be the first thing you would do to start your planning?

Determine a theme? The number of projectors? Front or rear projection? The number of screens? The media you'll use?

Whatever your answer, it's neither right nor wrong. Every successful multi-image producer plans in his or her own way. Some start by envisioning the overall images they want to create on the screen. Then they work backwards to determine how they'll create these impressions. Other producers work in just the opposite way.

They start by determining what they can accomplish with their equipment. Then they create their programs around their hardware. Still other producers have other methods of planning — and that's just the point. There is no "right" way to plan; no approach is more valid than any other.

But there is a lesson you can learn from these producers. No matter how they work — from software to hardware or from hardware to software — **they all have a planning process.**

And that's what you have to develop — a production planning process that works for you. It's a process you have to develop using your own experiences and inclinations as your guide. That's because planning is an individual process. No two producers plan exactly alike. In fact, you'll probably find yourself varying from your own planning process with each new production you begin. Why? Because each new production brings with it a new set of production requirements, and for the most part you won't be able to meet those requirements by duplicating exactly what you've done in the past. So, don't think of planning as a mechanical process — a set of instructions to follow. Since it's a dynamic process, your planning must allow for variations — in presentation requirements as well as in human attitudes and aptitudes. You must plan, but do it in a way that works for **you.**

Why bring this subject up? Because this section examines the factors that go into multi-image planning, and like all written material it's presented in linear fashion — a first part, a second, a third, and so on. This sort of sequential description suggests a formula approach. The impression is erroneous but unavoidable. It's just the way this book has to be written.

So, as you start your own planning, don't feel constrained by the sequential elements that we've listed. They're in a certain order, but you should **use** them in whatever sequence seems best for **you.**

Selecting Your Media

If you were an architect designing an office building, one of your crucial decisions would be to select the materials to be used during construction. In making this decision, you'd have to consider the function and purpose of the building, the use it would receive, its location, and the amount of money available for construction. On the basis of these considerations, you'd choose the building materials that best suited your purposes.

This kind of thinking also must be evident when you plan for a multi-image presentation. You must consider the function of the show, its projected use, the location or locations where it will be shown, the amount of money budgeted for production. With these factors considered, you must then choose the media that will best enable you to meet your goal.

Fortunately for you, the list of media you have to choose from is a lot shorter than an architect's catalogue of building materials. For practical purposes, the media you'll be working with can be divided into five categories: slides, motion pictures, light and other visual displays, music and sound effects, and live narrators and demonstrators.

1. Slides

Color transparency slides are the basic medium of multi-image production. In fact, you would be hard-pressed to find professionally produced multi-image presentations that didn't use slides for at least 80 percent of their visual content.

The common designation "2 by 2-inch slide" refers to the outer dimensions of standard slide mounts. (The 2 by 2-inch mount is considered standard because it's the size accepted by commonly used slide projectors.) The dimensions of the slide mount opening or aperture are determined by the film used to shoot the slides.

A number of factors—among them the location and use of and the budget for a presentation—must be considered when selecting presentation media.

The **135 slide**—or as it's more commonly called, the 35 mm slide—is the workhorse of multi-image production. Professional multi-image producers cite several reasons for the popularity of the format. First, 35 mm film is available in more film emulsions and speeds than any other format, so the multi-image producer has more flexibility when shooting.

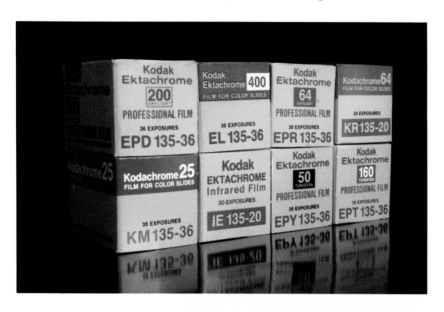

A second reason 35 mm film is so popular is that it's less costly to buy and process than other color-reversal transparency film formats. The difference in the average purchase price between a 12-exposure roll of 120 film and a 20-exposure roll of 35 mm film is minimal. But if a producer buys 36-exposure rolls of 35 mm film—as most do—then the savings in film cost is significant—about 30 percent per roll. When you consider that a half-hour presentation using nine projectors may require the shooting of almost 6,000 exposures—or 160 rolls—of 36-exposure film, you see that the savings realized through the use of 35 mm film can amount to hundreds of dollars.

Even greater savings are realized in processing. Again, using average costs for processing color reversal films, you find that the per-exposure cost for 36-exposure 35 mm slide processing is about 60 percent less than the cost for processing 12-exposure rolls of 120-size film. Multiply that savings by the number of rolls of film processed for a typical multi-image production and you have additional hundreds of dollars for other aspects of production.

Further savings are possible when using a 35 mm camera that accepts 100-foot (30.5 m) rolls of film. This length is equivalent to about 800 slides. The cost of shooting and processing slides in these quantities is about 50 percent less, per frame, than with a 36-exposure roll (for unmounted transparency strips).

Another appeal of the 35 mm format is that it's more convenient to use—especially during the editing stage of production. When 35 mm slides are processed, they're usually mounted in a standard 2 by 2-inch mount. They can be dropped into a slide tray and used "as is." On the other hand, 2¼ by 2¼-inch (57 x 57 mm) transparencies are normally returned from processing in strips, or mounted in a 2¾ by 2¾-inch (69.9 x 69.9 mm) mount. (The 2¼-inch square transparencies also can be returned mounted as super-slides, with a standard amount trimmed off all four sides of the image. This is a regular service offered by Eastman Kodak Company and other film processors. Because the original image must be trimmed, care must be taken during shooting to ensure nothing essential will be lost in the trimming process.)

A fourth reason most producers choose 35 mm film for production is that it can be used in most standard professional copy cameras and animation cameras without modifying the equipment. This means greater precision in registering and shooting multiple graphic or art slides. It also offers another convenience to a producer: Slides shot with this equipment will conform with all other slides shot for the production.

The 35 mm format does have one drawback. In certain circumstances the image area of the 35 mm slide can limit a producer's ability to create special effects or to project large images without using high-intensity lamps. When these circumstances arise, producers usually turn to super-slides.

Super-slides are just what their name implies—slides that offer more image area than any other slide format. But what makes them truly "super" to a multi-image producer is the fact that their large size yields up to 84 percent more area through which light is projected onto a screen as compared with a 35 mm slide. This extra light isn't always important, but it can be critical with large projected image sizes or in high ambient-light environments.

The large size of the super-slide does restrict its usefulness, however. As already mentioned, super-slides cost more than 35 mm slides to shoot, process, and mount.

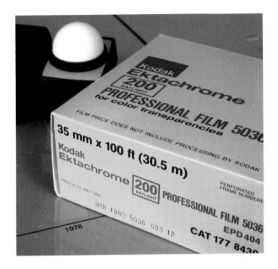

Super-slides were used to create this rear-projection presentation.

Some producers try to avoid the greater expense involved in using 120 film by shooting 35 mm slides and then enlarging them to super-slide dimensions. This saves some money—original film and processing costs are less. But these savings are almost totally offset when the 35 mm image is enlarged. This process requires the 35 mm slide to be enlarged to about 1.7 times its original size, then cut to fit the super-slide format. This process is not only expensive, it's also hard to obtain. It's a special service, one you may be able to get only from an out-of-town laboratory. It's also costly in another way: A large portion of a 35 mm slide's horizontal dimension is lost when it's cropped to fit the super-slide mount. And as with most enlargements of this magnitude, some image sharpness is lost. Of course, there's also the additional cost of trimming and mounting the resulting slide—a process that must be performed by hand.

Other 2 by 2-inch slide formats can be used for multi-image production, although their use is usually limited to non-professional applications.

A variety of slide formats and film emulsions can lead to varied results unless the slides are color corrected and duped.

The **126** format is produced for cameras such as the Kodak Instamatic® camera. Because of its smaller image area, the 126 format doesn't find broad application in professional multi-image presentations. It's used mainly in educational applications, particularly where presentations are produced using slides shot on a copy stand such as the Kodak Ektagraphic EF visualmaker.

Slides shot on 127, 620, 120, or 110 film, using nonprofessional cameras, can be used in 2 by 2-inch mounts. Here again, however, the use of these formats is usually restricted to situations in which budgetary or equipment limitations exist or in which image quality is not of critical importance.

Originals or Duplicates? This question has two aspects. The first is a simple case of economics—either the money or time needed to shoot original slides is in too short supply. In these cases, you'll probably have no choice but to go with dupes of existing slides.

You should, however, be prepared for a problem. Existing slides, by their very nature, are a collection of differences—different ages, different film emulsions, different mounts. And when these slides are projected, they're going to show their variations. Each slide will react to the heat of the projection lamp differently, creating focus problems and easily observable differences in photographic quality—and that's not what you want in your presentation.

The way to solve this problem—if you must use dupes of a mixture of existing slides—is to have them color corrected, if necessary, and duped using Kodak Ektachrome slide duplicating film 5071 (Process E-6), or equivalent. This will retain the visual quality of the slides while also providing a uniform film base. The resulting slides may not all look like originals, but they'll have as much visual and physical uniformity as is possible.

There's another problem inherent in using existing slides—and it's one that can't be solved by duping or any other means. That's the problem of inconsistent photographic style. Rarely will you find a single photographer who has shot all the existing slides you need. More likely, you'll be using slides shot by a number of photographers, each with a different level of know-how and experience and a highly personal photographic style. And you—and your audiences—will see the differences, almost as easily as spotting the differences in a page of signatures.

The way to solve this problem, of course, is to avoid it. Unless you're forced by circumstances into using dupes of existing slides, don't give the question of "original versus duplicate" a second thought. Go for original photography, shot for the show you're planning.

The second aspect of the original or dupe question is somewhat more involved. Basically, it's whether to use camera original slides or duplicates for the actual presentation. Certainly, budget will play a role here, as will time. If your presentation is scheduled for a once-only showing (but how often have you heard that?) or even a limited life, you would be better off using original slides. Being originals, they'll have good contrast, the color will generally be good, and you'll avoid the cost of duping.

On the other hand, your particular situation may well justify—or even demand—that you use dupes for the actual show. The most obvious instance of this is a show that would be used in a number of locations. Multiple sets would be a must. If a show is to have extended use, you'll probably need duplicate slides. This allows you to preserve the original set so that you can replace the presentation set when it gets worn or begins to fade. Another reason for duping is to create optical effects. You also could decide to correct a too broad exposure range or color balance through duping.

In any situation where it would be more desirable to use duplicate slides for your presentation, you must give careful consideration to the type of film emulsion to be used for shooting your originals. Some films, such as Kodachrome 25 and Kodachrome 64 film, produce very saturated colors and extreme sharpness, but they are also more difficult to duplicate without showing increased contrast. Other films, such as the Kodak Ektachrome films (Process E-6), have less contrast in the originals and are thus better suited for producing high-quality dupes.

The development of Ektachrome films and duping stocks for Process E-6 has created an additional advantage for using dupes of originals in your presentation. Dupes made using this combination of film stocks are almost impossible to distinguish from originals, even if the dupe and original are projected side by side. This being the case, using dupes gives you greater protection against the loss or damage of original slides.

Kodak Ektachrome films provide a range of speeds and are available for use in daylight or with tungsten light.

Kodak Ektachrome slide duplicating film 5071 (Process E-6) provides a convenient way to duplicate either original slides or an assortment of existing slides.

2. Motion Pictures

The introduction of motion pictures into a multi-image presentation adds an exciting dimension, but it can also add headaches. For many years, the only reliable method of starting and stopping a motion picture projector on cue was to have a projectionist physically controlling it. And in those cases, you could be assured that somewhere sat a producer with his or her fingers crossed.

Some of the earliest programmers offered the capability of remotely turning a motion picture projector on or off. Unfortunately, if there were many film sequences, and if the cues were relatively tight, it was still difficult to stop and recue the projector for the next sequence. Few projectors run at uniformly constant speeds and they have a tendency to "coast" inconsistently after the power is shut off. So, regardless of programmer capability, someone still has to recue the projector manually.

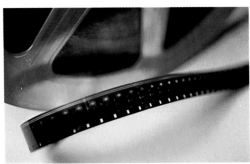

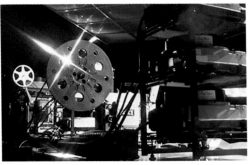

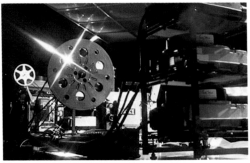

In recent years, however, a number of devices and techniques have been developed to make the use of motion pictures less traumatic for the producer. The more basic systems use devices mounted on the projector to sense foil tabs or notches on the film and turn the projector off at the right moment. More recent equipment (also more expensive) uses signals on the program's audiotape to maintain accurate synchronization with the film. A continuously variable stepping motor in the projector then controls the projector's speed. An added advantage of this type of equipment is that the sound—even lip sync—can be provided for the movie by the control tape. This eliminates the difference in sound quality that may otherwise be noticeable between the sound-on-tape for slide sequences and the sound-on-film for motion picture sequences.

Because of these developments, motion pictures are becoming a more important element in multi-image presentations. For the most part, the film sequences are used for emphasis. Like an italicized word in the middle of a sentence, motion picture film says "pay attention." This makes film an important tool in the multi-image producer's workshop.

16 mm film ranks as the number one motion picture format used by multi-image producers. This ranking reflects the professional respect given to 16 mm film as both a production and projection vehicle. Its users rely on it for several reasons:

- The equipment needed to shoot 16 mm footage is readily available and convenient to use. Even if an organization doesn't own a 16 mm camera, it can easily rent one from a motion picture equipment rental house. If the film sequences are not too complicated, an experienced still photographer can usually operate the camera. For more sophisticated productions, an organization can hire an experienced camera operator.

Motion pictures add another dimension to multi-image presentations.

- It's less expensive to shoot and process than 35 mm motion picture film. This is not only a factor of film size, but also one of the amount of film needed to shoot sequences. Using 16 mm film, a producer can shoot for a minute and use only 36 feet (11 m) of film. With 35 mm film, however, the same shooting time requires 90 feet (27.4 m) of film — or 250 percent more length, and more than twice the width.
- It's less expensive to produce release prints in 16 mm than in 35 mm.
- A 16 mm film crew requires fewer members than does a 35 mm crew. Although production techniques are not significantly different, 35 mm equipment is larger, heavier, and more unwieldy than 16 mm equipment and requires more crew members to move and operate it.
- By using 16 mm film, a producer has the choice of an optical or magnetic sound track. Some producers feel that a magnetic sound track on film will more closely approximate the sound on the ¼-inch audiotape used for the balance of a multi-image show. Also, if the film is not going to have its own sound track, the magnetic track can carry cues to actuate slides or even to shut the projector off.
- The light output of a 16 mm film projector and a 2 by 2-inch slide projector are roughly the same. So, if you're projecting images that are about the same size, there normally won't be a jarring difference in brightness on the screen.

35 mm film is the format to choose if extremely high-quality film images are an unalterable requirement. But you have to be able to afford them. And like the prospective yacht buyer, you probably can't afford this luxury if you have to ask about the price.

This format is the medium of top professional film producers. It's Hollywood's medium. It's Madison Avenue's medium. And if you've ever studied the budgets coming from these two production centers, you know 35 mm production needs financial muscle behind it.

There are also larger motion picture formats, but generally they are used only for theatrical productions and a few of the very high-budget multi-image productions. For this reason, their use will not be considered in this book.

Super 8 film isn't a big-screen format, so normally it isn't considered for multi-image presentations. But its small size — and especially the compactness and cooler operating temperatures of super 8 projectors — makes it a perfect format to use when producing multi-image presentations that will be contained in exhibits or displays. With these smaller image sizes a super 8 image can be projected with little loss of brightness.

Super 8 films can be produced using super 8 cameras and production equipment. Many professional producers prefer, however, to shoot and edit their original footage on 16 mm film, then reduce the film to super 8 when prints are made. This technique offers both high-quality production and low-cost distribution.

3. Light and Other Visual Displays

This is an area where the only limit on what you can do will be your imagination. Even the sky's not the limit any longer, because you could use a multi-image programmer to launch a rocket or send an activating signal to a communications satellite.

Special lighting effects, visual displays and lasers—such as those used to create this tunnel for a presentation introducing the Mazda RX-7—are other elements that can be considered when planning a presentation. (Photo: Laurence Deutsch Design, Inc.)

If this sounds like an exaggerated claim, just consider what a programmer is and what it does. The more sophisticated of the programmers on the market today are mini-computers that can operate **anything** controlled electronically or powered electrically. On a practical level, these programmers allow you to use a wide variety of visual media in your presentation. Spotlights. Strip lights. Turntables. Conveyors. Laser projections. Holographic projections. You can open and shut screens, draperies, curtains, doors, and skylights. You can turn on fountains, raise a flag, or lower a background.

Most producers use visual displays to get or direct attention, to generate interest and excitement, and to maintain enthusiasm and a sense of anticipation during a long presentation. Earlier, motion pictures were likened to an italicized word in a sentence. To continue this analogy, light and visual displays are the equivalent of exclamation points. They say "excitement!" This emotional ingredient adds impact to the message of your multi-image presentation.

4. Music and Sound Effects

It's often easy for a multi-image producer to forget that **music and sound effects may be just as important as visual effects in the success of a presentation.** This usually happens because most producers begin as photographers or cinematographers, not as sound specialists. Their visual bias is well established.

If you find yourself in this category, take time to consider the types of sound you'll use in your presentation. You'll be working with two basic sources of sound—original sound and canned sound.

Original sound refers to an original musical composition (or an adaptation of a copyrighted composition), created for a specific presentation. It can also refer to background sounds recorded at the site of a photographic session.

"**Canned**" sound refers to the music and sound effects previously recorded and contained on records produced by companies that specialize in sound libraries.

Of these two categories, canned or library sound is the easiest to use and the least expensive. Descriptions of the music and sound effects contained on the records are printed in library catalogues and on the jackets of the records themselves. Working with these descriptions will require some time on your part because a library producer's description of a record's contents may not agree with your evaluation. So, some trial-and-error searching will be required. To pay for the music or effects taken from a library record, you keep track of the number of "needle-drops" — or separate segments — recorded from each record. You then pay the library company a royalty for each needle drop.

One drawback in using library music and effects is that the selections are, to some degree, stereotyped. The music is composed, arranged, and played so that it can be used by anyone, from the producer of an industrial film to a retail advertiser creating a local radio spot. Its basic appeal is its universality — that's how the library companies make their money. The same is true for "canned" sound effects. Think of the sound used to simulate a car crash on radio dramas. Over the years it became a cliché — the screeching brakes, the crunching metal, the shattering glass — until it finally became a basic ingredient in many comedy routines. This type of sound effect won't cost you much to use, but in your planning you must remember that library sound effects are general approximations, not the specific sound of a specific event.

This drawback can be overcome with the use of original sound and music. But like all original creations, you must expect to pay a little more to obtain them. Recording on-location sound for your presentation is not that difficult; and it's worthwhile because it provides the audience with a sense of realism. You need a sound re-

cording specialist and professional equipment; later you'll also need a sound editor. This, for the most part, is usually a talent expense — you'll pay for a sound specialist who'll provide the equipment. But if you plan to use sound to carry an important part of your presentation's message, your money will be well spent.

Hiring a composer to create or adapt an original score for your presentation is another matter. If the composer has experience composing for movies or multi-image presentations, you can expect to pay for this know-how. John Williams, who has composed symphonic works as well as the music for **Jaws, Star Wars,** and **Close Encounters of the Third Kind,** said composing for film is far more difficult than composing without any time or thematic limitations. And that's really what you're asking a composer to do when you ask for an original score: to create a musical mood that emotionally involves your audience. You're asking the composer to interpret your theme musically and to do it within the time span of your production sequences. So when do you plan to use original music? When the objective justifies a budget sufficient to pay for it.

Library music, which is paid for by the "needle drop," is the most economical way to create a music track for a presentation.

55

5. Live Narrator and Demonstrators

For the most part, **you** don't plan to use a live narrator or demonstrator. This decision is made much earlier, when the presentation requirements are established with the client. But when production begins, the appearance of narrators or demonstrators in a show does present two planning questions: How do you use them? And what specific problems do they create?

Using a narrator or demonstrator in a multi-image presentation creates problems similar to those a stage director faces when laying out stage movements for actors. In both cases, the goal is to avoid the static—a "talking head" or a collection of stick figures waiting for a puppeteer to pull their strings. In short, the people you put into your presentation should look and act as if they belong there.

Achieving this goal is easier with demonstrators. They can be directed to come on stage, demonstrate a product or activity, and then leave. Spotlights can be used to lead the demonstrators on and off stage. They can appear from behind special screens. But whatever the technique used, you must plan to **incorporate** these people into the flow of a presentation.

Working with a narrator or narrators is a bit more difficult. Your major problem is to give attention to both the narrator and the multi-image presentation without detracting from either. To do this, you must make the narrator essential to what's happening on the screen, not just a neutral commentator. In some situations, the narrator's role, title, or organizational position will establish this essential connection. For example, using the president of a company to narrate a presentation of corporate accomplishments and goals creates an essential connection. On the other hand, it's often desirable to present a company executive on film—especially if he or she is an inexperienced narrator. The footage can then be edited to effectively convey the message.

A slightly different situation exists when multi-image presentations share the billing with live performers. Rock musicians have used light shows and multi-image presentations at concerts since the late 1960s. Since then other performers, both musical and dramatic, have borrowed and adapted the idea, replacing static stage backgrounds with multi-image presentations.

Sometimes the presentation must carry information essential to an audience's total understanding of the performance. The actor Leonard Nimoy, for example, has written and produced a one-man show examining the life of Vincent Van Gogh. Nimoy portrays Van Gogh's younger brother, Theo. As he examines Van Gogh's life and work, often reading from letters written by Vincent to Theo, a two-screen, multi-image presentation supplies the corresponding visual details of Van Gogh's life.

Multi-image presentations are often used to establish the visual background and mood for live performances. (Photo: Laurence Deutsch Design, Inc.)

Sometimes the presentation communicates a more subjective, emotional mood. Producers from Laurence Deutsch Design, Inc. were asked to do this for a symphonic concert presented at the Waikiki Shell in Honolulu. The performance featured music from the sound tracks of **Star Wars** and **2001: A Space Odyssey** and from symphonic compositions such as **The Planets.** On stage, live dancers performed against a kaleidoscopic background created using lasers and a large-screen, multi-image presentation.

Although the situation in this type of presentation is different than that when a narrator reads a multi-image script, the producer's problem is essentially the same: He or she must integrate both elements into the flow of the performance. Performers and multi-image presentation must contribute to a common communications goal, not compete against one another for an audience's attention.

Working with live performers creates one additional problem for a multi-image producer. When production is completed, but prior to audience presentations, the performers must be available for rehearsals. Not rehearsal, but rehearsals. The producer must be able to see how the performers move and interact with the visual presentation. This takes time—for the performers to learn their movements and lines, and for the producer to evaluate and refine their performances. And as already discussed, time in multi-image production equals money.

Selecting The Visual Format

One of the first acts an oil painter performs before mixing colors or selecting a brush is to prepare the canvas. Guided by an inner image of the eventual work, the artist searches for or constructs a frame with a scale to match his or her intentions. The artist is deliberate in determining shape and dimensions, because this frame provides the broad outline of the final painting. Only when the painter is satisfied with the frame does he or she put on the canvas. And only then is he or she ready to paint.

The multi-image producer must take similar, deliberate care in preparing the "canvas" for his or her presentation. The producer's canvas, of course, is the screen; the frame he or she constructs for this screen is called the visual format.

The visual format prescribes the overall "look" for a multi-image presentation. It has nothing to do with the visual content of a show. In fact, it's not uncommon for a producer to determine a presentation's visual format before giving any thought to the visual images he or she will eventually use.

The visual format is the frame, the outline, the grid system, within which the visuals will appear. As such it serves two purposes. In one sense, the visual format is similar in function to newspaper layout sheets. These sheets, which duplicate the overall dimensions and column divisions of a newspaper, are used by editors to plan the placement of headlines, stories, photographs and advertising. The layout sheet is a spatial organizer, and to a certain extent, so is the visual format of a multi-image presentation.

But the visual format is also much more than a mechanical planning device. Conceived creatively, the visual format contributes to the expression of a multi-image presentation's message. An excellent example of the creative contribution made by a visual format can be seen on page 108. The visual format for the five-screen presentation created by Laurence Deutsch Design for the Mattel Toy Company echoes the show's theme — "Dimension '78." By creating the illusion of depth — or dimension — the visual format reinforces the show's visual and thematic message.

Creating The Format

Creating a visual format requires you to juggle four factors — audience size, room dimensions, screen size, and the placement of your projectors and equipment. In your planning, you move back and forth, from one factor to another, as you narrow your various possibilities to the one format that works best — spatially and aesthetically — with your presentation. In the text that follows, you'll find these factors presented separately and sequentially, as if one factor is to be resolved before moving to the next. However, in practice, this approach is not only impractical, it's also impossible, as this section will indicate.

To emphasize the concept of "dimension" for the Mattel Toy Company's **Dimension '78** presentation, Laurence Deutsch Design created a visual format that seems to emerge from the center screen. (Photo: Laurence Deutsch Design, Inc.)

Audience Size. This should be the easiest factor for you to resolve. In fact, you may have little to say about it. The number of people who will see your presentation may have been set by someone else during the initial planning for the presentation. If this is the case, you have a number to work with. Your only other question is whether these people will view the presentation in one showing or over a number of showings.

If the audience size is unknown, as it could be for a convention or trade show presentation where the performance is "open to the public," your problem becomes more complex. You, in effect, have to limit the audience size to the number of people the presentation site will seat comfortably in full view of the screen. But you can't do this until you know something about the room or theatre where the presentation will be shown. So you must keep this "ball" in the air for a while.

Important factors to consider when developing a visual format are the number of people who are to see the show and the dimensions of the presentation site. (Photo: Laurence Deutsch Design, Inc.)

Room Size. Here again, obtaining the information on room size isn't difficult. Working with it, however, requires you to continue your juggling act with all the other factors.

The easiest way to get accurate measurements of the presentation site is to ask the people responsible for the physical management of the facility. This may be the facilities management staff of your company; or it may be the head of maintenance, the events coordinator or banquet manager at the hotel or convention hall where you'll put on your show. When you talk with these people, tell them you're interested in four measurements— length, width, floor-to-ceiling height, and the dimensions and location of any object such as a raised platform, air duct, column or chandelier that might interfere with projection.

The length and width measurements will enable you to calculate the amount of floor space you have for seats, aisles, screens, and projection equipment. The floor-to-ceiling measurement— and the size and location of any obstructions— can be used to calculate your maximum screen height. Keep in mind that if you use front-screen projection, the bottom edge of your screen should be no less than four feet off the floor if you are to project images that will clear the heads of a seated audience. Whether or not you'll be able to use the maximum height available will depend on the placement of your projectors, so this "ball" remains in the air for a while too.

Actual Screen Size and the Placement of Projectors. The actual size of your image on the screen is a factor of projection distance and the focal length of the lenses on your projectors. The table on page 61 gives you the relationship between distance, focal lengths, and image size.

For example, if you're using 35 mm slides and your maximum screen height is four feet and you want to use the full height, you look along the horizontal listing for 48 inches (1.2 m). There you'll find figures representing projection distances using the focal-length lenses listed. For example, if you're using projectors with seven-inch lenses, your projection distance is about 32½ feet (9.9 m).

It would be simple if you could just pick the projection distance you wanted, then use the focal-length lens called for. But several other factors enter into the decision.

First, if you're planning to use front projection, the screen-to-projector distance may place your projection booth in the middle of your audience. This could eliminate valuable seating and obstruct the view of people to the side and rear of the booth.

Second, the different focal-length lenses, while mechanically interchangeable, are not necessarily interchangeable in terms of the level of illumination on the screen. As the intensity of the light projected to the screen is reduced, so is the brilliance of the images on the screen.

Third, the optics in lenses with short or long focal lengths are more costly than those in a standard lens. These lenses can cost as much as $300 to $500 each, and can become a major expenditure for a multi-projector show.

So choosing projection distance is going to require some juggling of its own. You'll need to consider the desired height and width of your images, the size and seating of your audience, and the projection distance you have to—or want to—work with. You also must decide between front-screen and rear-screen projection.

Projection Distance Table for
Kodak Ektagraphic Slide Projectors
(Projection distances are approximate and are measured from projector gate to screen.)

Nominal Aperture Dimensions* of 2 x 2-Inch Slide Mounts

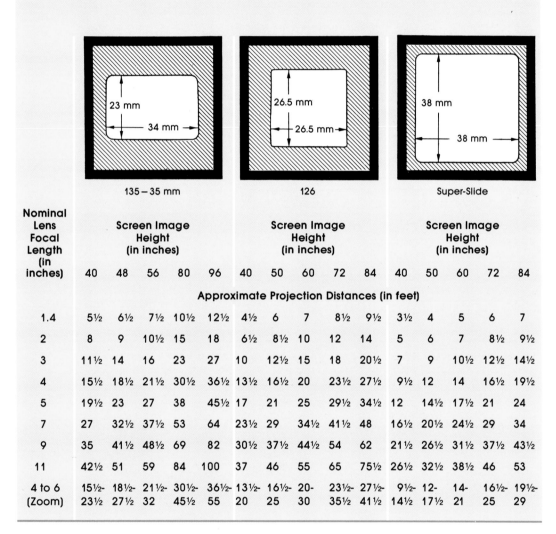

| 135 – 35 mm | 126 | Super-Slide |

Nominal Lens Focal Length (in inches)	Screen Image Height (in inches)					Screen Image Height (in inches)					Screen Image Height (in inches)				
	40	48	56	80	96	40	50	60	72	84	40	50	60	72	84
Approximate Projection Distances (in feet)															
1.4	5½	6½	7½	10½	12½	4½	6	7	8½	9½	3½	4	5	6	7
2	8	9	10½	15	18	6½	8½	10	12	14	5	6	7	8½	9½
3	11½	14	16	23	27	10	12½	15	18	20½	7	9	10½	12½	14½
4	15½	18½	21½	30½	36½	13½	16½	20	23½	27½	9½	12	14	16½	19½
5	19½	23	27	38	45½	17	21	25	29½	34½	12	14½	17½	21	24
7	27	32½	37½	53	64	23½	29	34½	41½	48	16½	20½	24½	29	34
9	35	41½	48½	69	82	30½	37½	44½	54	62	21½	26½	31½	37½	43½
11	42½	51	59	84	100	37	46	55	65	75½	26½	32½	38½	46	53
4 to 6 (Zoom)	15½-23½	18½-27½	21½-32	30½-45½	36½-55	13½-20	16½-25	20-30	23½-35½	27½-41½	9½-14½	12-17½	14-21	16½-25	19½-29

Note: Kodak supplies a 2½-inch **Ektanar** Projection Lens (not shown). It is designed especially for use in study carrels and in small rear-projection cabinets. With a standard 35 mm slide, the lens produces an 8-inch-wide image in a 26-inch distance (back of projector to the front of the screen). Image sizes over 12 inches wide are not recommended.

*Dimension tolerances may vary with mounts of different sizes and manufacture.

Front Projection Vs Rear Projection

Choosing between these two projection formats is not a simple matter of flipping a coin. But neither is it always a strictly objective matter where specific sets of circumstances dictate the use of one approach over the other. As with all the creative decisions you'll have to make, choosing a projection format will require sensitivity to the **physical** as well as the aesthetic requirements of your presentation.

Front-screen projection usually offers greater quality of projected images. They're brighter, with more even illumination, than most rear-projected images. And contrast is better because the image is projected onto a screen, not through it. There is, however, one quali-fication to this statement. If in trying to avoid the placement of a projection booth in the middle of an audience you decide to use long focal-length lenses, you could encounter a problem of reduced light intensity.

Rear-screen projection equipment is set up behind the screen, out of sight of the audience. The images are projected onto a translucent screen (for viewing from the other side). With this arrange-ment there's no booth to interfere with the openness of the seating area. And because the projection equipment is out of sight, there's no need to spend time or money to drape it.

Some people also feel rear projection offers the advantage of reduced projection distance. While it is true that projection distances can be somewhat shorter—because you can move the projectors closer to the screen—this is often accomplished through a trade-off in image quality. That trade-off is a product of the projected image's "bend angle"—or the angle at which the light from the projector must bend to reach the viewer's eyes. (See diagram.)

To correct this problem, the projection equipment must be moved away from the screen. To do this, however, nullifies the so-called advantage of short projection distance. (You can also "fold" your projected images using mirrors. See diagram for one method.) Rear projection has one major disadvantage. Its images, from light projected **through** a screen, are generally less sharp, less bright, and have less contrast than those of front-screen projection.

Rear-screen projection is also used—and used extensively—in multi-image presentations for displays and exhibits. In these applications, the projectors are usually contained within an exhibit structure. The images projected are usually small, and very bright, so lenses with focal lengths as short as one inch often can be used without creating dark areas from the "bend angle" problems connected with larger images.

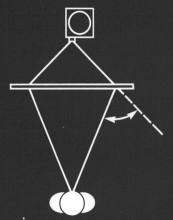

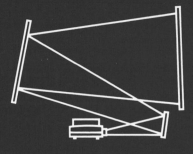

Nailing Down Some Decisions

You've been juggling quite a few numbers and considerations. The time has now come to take these considerations out of the air and put them down on paper.

The first decision you must make is projection distance. You've examined the size of your presentation site, the probable size of your audience, and the size you desire for your visual format. You've also examined the benefits and drawbacks of both front-screen and rear-screen projection.

So where are you going to put your projectors? Don't look for any additional help or suggestions from these pages. There aren't any. There can't be. That's because the decision on where to place your projectors is unique to the presentation you're planning. You've read all the guidelines and studied the room and seating arrangements. You have an idea of the eventual height you want for your images. So, all things considered, where are you going to put your projectors—in front of the screen or behind it?

Once you've made the decision, go back to the Projection Distance Table and determine the actual height you'll use in your visual format. For example, if you're using 35 mm slides and your projection distance is 50 feet (15.2 m), then the actual image height would be 75 inches (1.9 m) using seven-inch focal-length lenses.

Determining Image Width. Now that you have the actual height of the screen, you can **choose** a suitable width. The emphasis on the word "choose" means you don't fall back solely on a formula to determine screen width. Aesthetic considerations—your creative judgment—enter the decision too.

But you start with a simple formula based on the image dimensions—or aspect ratio—of your slides. The aspect ratio of a slide expresses the relationship between its height and width. A standard mounted horizontal 35 mm slide, for example, is nominally 24 mm high by 36 mm wide. Divide the height by the width and you have an aspect ratio of 1 to 1.5 (1:1.5)—or as it's more commonly referred to, 2:3. (This notation is used to avoid fractional numbers.) The aspect ratio of a square super-slide is, of course, 1:1.

Using these ratios in conjunction with your figure for actual screen height allows you to set the screen width. For example, if the actual height of your screen is 80 inches (2 m) and you're projecting 35 mm slides, the minimum width you need **for a single image** is 120 inches (3 m)—1.5 times the height.

It's at this point that your creativity comes into play. You know the width you need for a single image on the screen. So you ask yourself some questions: How many images do you want to put on the screen at once? What size slides do you want to use? Will you use a combination of slide formats? As you answer these questions, you begin to develop your overall visual format and dimensions.

The illustrations on the following pages show some of the more common screen formats used by professional producers.

One of the most common multi-image screen formats uses two 2:3 ratio images side by side. The presentation requirements for a production will determine the number of projectors used in this format.

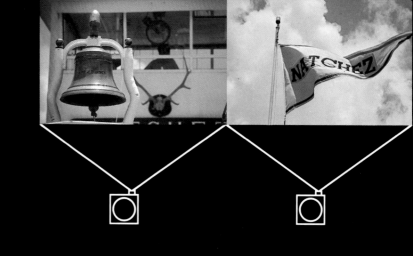

The same screen format as above, only in this case illustrated to show the use of 1:1 ratio super-slides.

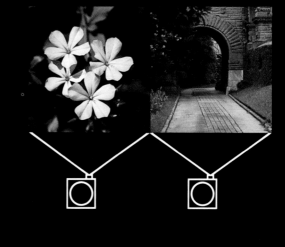

One step further in sophistication: A screen format using three 2:3 ratio images side by side. Here again, presentation requirements for a production will determine the number of projectors used in this format, as well as the number used for each screen area.

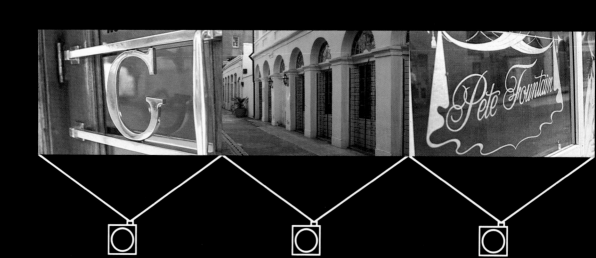

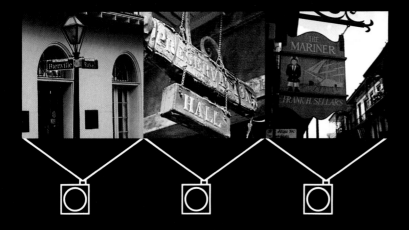

A super-slide variation of the 2:3 ratio, three-screen format: Note that the overall dimensions of the screen are in the same ratio as that for the two-projector format illustrated at the far left.

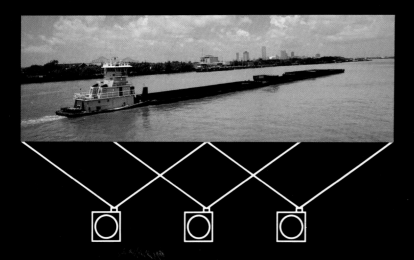

A more sophisticated format using overlapping images: The projection setup consists of two 2:3 ratio images side by side plus an overlapping 2:3 ratio image focused on the center of the screen. This setup allows a producer to create full-screen panoramas as well as to create sophisticated animation. The format illustrated below is a similar overlapping setup using five projector banks. Three of the projector banks are aimed at three 2:3 ratio screen areas (similar to the format illustrated at the far left). Two additional projector banks are aimed at areas that overlap the basic screen areas.

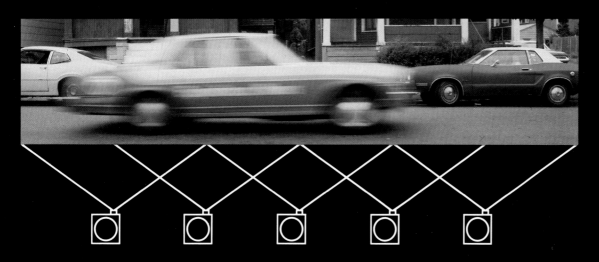

A variation of the format using three overlapping image areas: In this setup, images projected by the left and right banks of projectors can be masked to create vertical half-frame images (1:1½ ratio), while the center bank of projectors uses 2:3 ratio images.

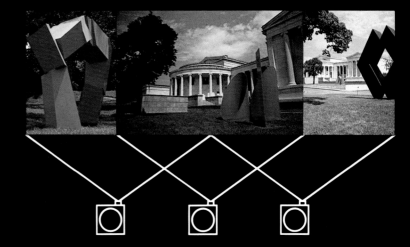

Another variation on the above, with all images masked to vertical half-frame (1:1½ ratio) formats: This format is particularly useful when the subject of the presentation presents a basically vertical orientation.

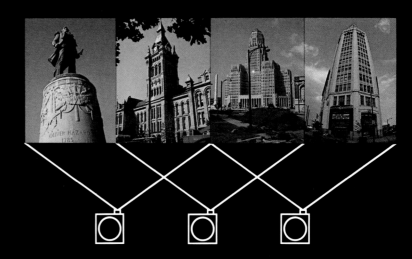

Still another variation of the format using three overlapping image areas: In this setup, quarter-frame images are projected into each of the screen areas. This format is frequently used when a producer wants to show many simultaneous images and when numerous image changes are required in the center screen area.

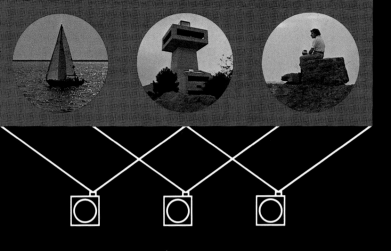

Creative masking allows a producer to vary the basic rectangular and square image formats.

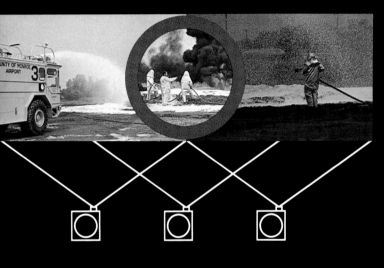

Another example of creative masking: The inner edges of the left and right slides are masked by semicircles; the center slide is masked as are the slides in the illustration above.

Placing an image within an image: The effect is created by blocking out an area in the larger, background slide through the use of a mask; the smaller image is then projected into the masked area.

Your efforts to determine the size, ratio, outer shape, and inner layout of your screen area will result in your visual format. It will be both the frame and the canvas for your presentation. The visual format you create will be determined by the factors already discussed **plus** your creative interpretation of your presentation's objective, theme, and requirements. The format you develop may be similar to the formats illustrated on the preceeding pages. Or it may be a "look" far from the common or the standard. But whatever format you create, it should grow out of the specific needs of a particular show.

There's a simple standard with which to judge your visual format:

Even if your format is a carbon copy of a format used in dozens of other shows, when you use it there should be only one reason: It's the best possible format in which to present your images.

Even if your format is unique—a never-before-used combination of sizes, shapes, and layouts—when you use it there should be only one reason: It's the best possible format in which to present your images.

If your format can't meet this simple standard, you're probably guilty of either unthinking imitation or undisciplined creativity.

Seating and the Visual Format. It would be convenient if your planning could stop at this point—with the overall format and dimensions of your screen. But to stop here would be to ignore the audience, which must be able to see **all** of the images on the screen. This consideration brings you back to audience size and seating. It may also mean some modifications in your proposed visual format.

To determine the number of people who will be able to see the screen from top to bottom, and from edge to edge, you have to turn to two familiar sets of dimensions—the size of the room and the size of your screen. With these figures perform the following calculations:

1. Multiply your actual screen height by two. The answer you get is the minimum distance you should allow between the screen and the front row of seats. If you reduce this distance, you're likely to have people in the front rows who can't see the entire presentation without craning their necks or turning their heads from side to side.

2. Multiply your actual screen height by eight. This figure represents the maximum distance you should allow between the screen and the last row of seats. Beyond this distance, images on the screen—especially those that contain words, numbers, or other details—may be too small to be seen distinctly.

These two calculations give you the maximum depth of your seating area.

3. Now take the width of the room and leave yourself three feet at either side for aisles. Also allow eight feet for two more aisles within the seating pattern. (Some producers elect to use one center aisle. The major disadvantage of this practice is that the aisle consumes what is the optimum viewing area.) Subtract the width of the aisles from the total width of the room.

Choosing The Screen Surface

Your choice of screen material will have some bearing on the number of people you'll be able to seat at your presentation. These differences are attributable to the different viewing characteristics for each screen.

Matte screens diffuse light evenly in all directions. Images on matte screens appear almost equally bright over a large viewing angle.

Lenticular screens may have regular or irregular patterns of stripes, ribs, rectangles, or diamond-shaped areas. The pattern is too small to see at viewing distances for which the screen is designed.

With most lenticular surfaces, light coming from above or either side is reflected downward or to the opposite side, giving the screen superior control of ambient light. Many lenticular screens also provide an image three or four times as bright as a matte screen.

Beaded screens are useful in long, narrow rooms or other locations where most viewers are near the projector beam. These screens have white surfaces covered with small, clear glass beads. Most of the light reaching the beads is reflected back toward its source. Because of this, beaded screens are not a good choice for a multi-image presentation, where projectors may be spread out over a wide area in the back of a room. A person sitting near a projector beam at one side of the room will see the image from that projector as being several times as bright as the image from a projector on the far side of the room.

Rear-Projection screens are translucent. They allow projected image light to pass through the screen. Images tend to have a softer appearance because of the thickness and diffusion characteristics of the screen material. The viewing angle of a rear-projection screen is not necessarily narrow. It depends on how much diffusion it has and the bend angle required by the projector and viewer locations. Also, image contrast is usually lower with rear-projection screens than with front-projection screens (in similarly darkened rooms).

4. Multiply the seating depth figure from Steps 1 and 2 by the width figure from Step 3. This will give you the maximum floor space usable for seating. (If you plan to use front projection, with the projection booth situated within the seating area, calculate the area of the booth and subtract it from the total seating area.)

5. In calculating your seating capacity, allow at least four square feet for each person. This figure includes space for seating and for a narrow aisle between rows of seats. Now divide the total area from Step 4 by four. Your answer equals the number of people you can comfortably fit into the viewing area.

6. The next calculation enables you to eliminate marginal seating—those seats that because of the acute angle of vision to the far edges of a screen would mean their occupants might see some images that would be distorted or shadowy (or not see them at all). Then you must determine what type of screen you'll be using. (See box at left for definitions of the most common types.) Then, from each side of the screen, measure a distance equal to one-third of the screen width. From these points on the screen, draw lines toward the side walls according to the following angles:

- 90° if you're using a matte screen for front projection
- 60° if you're using a lenticular screen for front projection
- 50° if you're using a beaded screen for front projection or one of the more popular rear projection screen materials

The lines you draw may eliminate some seating from the front corners of your seating pattern. Subtract these seats from your figure obtained in Step 5. The resulting pattern defines your optimum viewing area and sets the number of people who will be seated for each performance.

Note: The steps and calculations to be followed in creating your visual format are summarized in the Appendix on page 240.

Images projected onto a curved screen, from a presentation produced for Süddeutsche Zeitung, Munich.

Projecting Onto Curved Screens

A curved screen is one of the most inviting screen formats used by multi-image producers. Its curved surface seems to make a presentation site more intimate by drawing people into a closer relationship with the visuals. Its novelty also can give an otherwise straightforward presentation more interest and impact.

But unless it's used properly, a curved screen can also invite problems for the multi-image producer, with poor image focus and image degradation being the two most common.

To avoid these problems, you should consider the following factors when planning for a curved-screen presentation:

1. The radius of the curved screen should be about equal to the maximum viewing distance (although you can stretch this figure up to 1½ times the maximum viewing distance). Curvatures of this depth will not create focus problems with most projection lenses. If, for example, you were using a 7-inch f/3.5 lens to project an image 50 feet, you would have a depth of field of 5 to 10 feet.

2. Use a rough matte surface to reduce the possibility of image degradation caused by light reflecting from one point on the screen to another. A ribbed matte surface or a sand-plaster surface is best for reducing stray reflection.

3. To eliminate keystoning (and to simplify image focus), align your projectors (a) so each covers no more than 20 to 30 degrees of screen curvature and (b) so each projects at a right angle to the chord marking the outer edges of the image area. (See illustration.)

4. Projector beams **can** cross each other without degrading the projected image, although in some situations it's better not to cross them. Two simple rules of thumb should help you decide whether or not to cross projection beams.

If the point at which the beams cross is within the main part of the presentation site, the left projector should project the right image and the right projector should project the left image. (See illustration.)

If the point at which the beams cross is near the projectors (or at an imaginary point behind the projectors) it's usually preferable to have the right projector aimed at the right screen and the left projector at the left screen. (See illustration.)

These guidelines will help you with simple curved-screen formats. If you plan to project onto domes or 360-degree screens, you should seek the advice of an optical engineer.

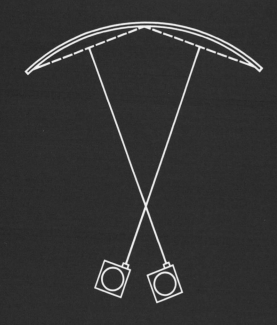

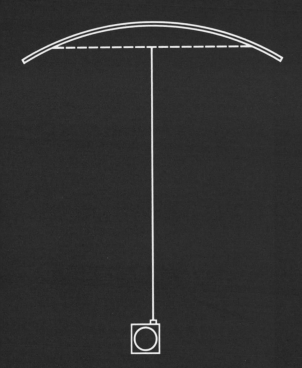

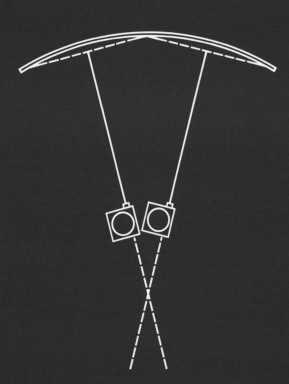

Even large screens can be extremely portable.

Buy, Rent, or Build?

Should you buy, rent, or build a screen for your multi-image presentation? There are no guidelines to use in answering this question; each presentation produces its own requirement. So, common sense is usually your best guide.

Some observations, however, may help you in planning.

- You can't consider renting unless your presentation is based on a standard screen format or ratio. Rental screens are usually available in the following ratios: 3:2, 2:1, and 3:1. A screen—or multiple of screens—based on these ratios can be rented from one of the companies listed in the appendix. (It's also possible to "dress" or mask a screen to a different shape by using strips of black velour material.)

Curved screens and other special formats must be custom built.

- All other screen ratios—and screens with special formats—are considered custom screens. They must be built to your specifications. But don't let the thought of having a screen built lead you to abandon the idea; the cost may not be as prohibitive as you think. Custom-built screens can be ordered from the companies listed in the appendix.

- Renting a screen is usually most economical when your presentation is going to be shown in one location over a period of a few days. It may also prove economical if the presentation is to be shown in several locations, but only occasionally, and not for an extended period of time.

- It's usually more economical to buy a screen when your presentation will be traveling to a number of locations or if it's going to be shown continuously at one location. With this type of use, you'll quickly reach the point where the cost of rentals would exceed the cost of purchase. (For portable use, stands, frames, and shipping cases for screens are also important considerations.)

- If your requirement is for a comparatively small front-projection screen, you might consider building your own. Any flat surface—plywood, a matte-finish plastic laminate like polystyrene mounting board, even stiff cardboard—can be used for a screen if it's free of surface markings. The limitations of this type of "homemade" screen are its size and the reflecting properties of its surface.

Planning For Your Programmer

It's time to slow down the planning process for a moment.

So far in this chapter you've examined the various media you can use in planning your multi-image presentation. And you've looked at the factors involved in designing a visual format.

Ideas—for the combination of media you'll use, for the screen format you'll create, maybe even for some of the visual sequences in your presentation—should be swirling through your imagination now. You should feel as if you could produce the most creative, lavish, exciting, and informative presentation yet seen on a multi-image screen, or the most practical, no-frills but entertaining presentation possible. After all, the message of this book is to untether your imagination and explore your creative possibilities.

But now it's time for your imagination to confront reality. And reality, in this case, is the capability of your programming equipment. In the previous pages you've been told that programming equipment—especially the new generation of computerized programming equipment—has been a liberating factor in multi-image production. This equipment allows producers to control equipment and create effects in ways never before possible.

That's true. But it's also true that your programming equipment is the single most limiting element in your planning. Your programming equipment may allow you to control 50 projectors. By the same token, it may also prevent you from using 51. Your equipment may allow you to program 150 slide changes a minute. But it also keeps you from achieving faster pacing. Your programmer may store 1,000 cues. But that limit may prevent you from creating more elaborate effects in a longer presentation.

So regardless of how advanced your programmer is, it's also your one constant in planning. It defines the possible. You know what media and combination of media are available to you. Your programmer determines what you can use. You know how to design a visual format to frame your multiple images. Your programmer determines the number of images you can place inside that frame. Beyond these points, your creative reach threatens to exceed your technological grasp.

What are your limits? Your programmer instructions will tell you. If you don't know already, take the time to find out. And be sure you know the practical meaning of each capacity. To read that a programmer offers "192 slides-per-minute maximum advance speed with a 4-projector dissolve" contributes little to your ability to plan unless you know what that capability allows you to do. You have to know—through use or experimentation—what opportunities and limitations each capability represents.

Once you know what your programmer can do, just keep in mind one caution: **Don't try to stretch the capabilities of your equipment.**

So plan to do what is possible to do—or more correctly, what is possible for your programmer to do. That's erecting a practical boundary, but it still leaves you with plenty of creative leeway.

Programmer capabilities are a basic consideration in planning.

Creating A Presentation In A Pit

How do you project multiple images into a hole 11 feet below the floor of a presentation site?

With mirrors, of course.

That's the illusion—and the technique—Eastman Kodak Company used to create a 12-projector multi-image presentation aptly named **The Image Pit.**

The presentation takes advantage of a basic characteristic of flat mirrors: They create an illusion of depth equal to the distance between a mirror and a mirrored subject. At **The Image Pit,** slides are projected onto the ceiling, but the audience views them on mirrors placed on the floor. The effect created is that of looking into a hole—a pit—containing images.

The Image Pit was conceived and developed to demonstrate that a major multi-image presentation could be shown in a small area. Kodak's communications objective for this presentation, to be used at conventions of audiovisual users, was "to excite people about the possibilities of multi-image presentations."

The company also had several specific presentation requirements:

- the presentation was to be "fast-paced and continuous;"
- it was "to eliminate the need to provide seating arrangements;"
- it was "to display Kodak Ektagraphic projectors prominently";
- and it was "to promote an even flow of traffic through the company's exhibit area so salespeople could have a chance to talk to people who had seen the presentation." (This last requirement was meant to solve the problem presented when a large audience leaves a major presentation and scatters.)

These requirements called for a presentation that people could walk up to, watch without having to wait for a specific starting time, and then leave having seen and heard the company's message. Thus, **The Image Pit** was born.

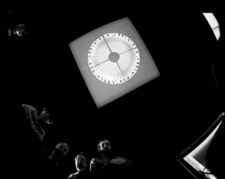

(above) Eastman Kodak Company's **Image Pit** as it appeared at the 1979 NAVA Convention. (top right) As viewers looked into the **Pit,** they saw images that appeared to be 11 to 12 feet below floor level. (bottom right) To create the effect, images were first projected onto the **Pit's** ceiling-mounted screen; viewers actually saw mirror-reflected images.

The **Image Pit** resembles a kiosk similar to those used at fairs and carnivals. Its 9-foot square, counter-like perimeter forms the viewing area for the presentation. Inside the structure, in a 3-foot by 3-foot area directly in the center, is a specially built projection frame containing 12 Ektagraphic projectors. Eight projectors are located on the bottom level of the frame, four on the upper level. Small front-surface mirrors of high optical quality are positioned in front of the lenses (4-inch, f/2.8) at an adjustable angle of about 45 degrees, to direct the projector beams upward.

On the floor surrounding the projectors are four 3-foot by 6-foot (.91 x 1.8 m) mirrors. The mirror edges are ground smooth.

About 8½ feet (2.6 m) above this structure is a 16-foot (4.9 m) square hood containing four loudspeakers and a matte projection screen about 8 feet square. But this isn't the screen viewers see. Instead, they look down into the kiosk, where they see the image reflected in the structure's floor-mounted mirrors. Because the images seen in the mirrors seem to be as far below the mirrors as they actually are above, the optical illusion created in the eyes of viewers is that of an image 11 to 12 feet (3.4 to 3.7 m) below the surface of the presentation site (about 16 feet from their eyes). (See diagram of sight-line analysis.) In addition, even though a viewer is looking into a mirror 3 feet across, the image seen appears to be about 8 feet square.

Because viewers can look into **The Image Pit** from four sides, special care had to be taken in shooting the slides for the five-minute presentation. Many of the slides taken for the presentation were shot from above the subjects; when projected, these images were oriented to viewers on all four sides. Graphic slides were shot so they could be rotated (through programming) during presentation; again, when projected, these slides achieved "right side up" orientation for viewers on all sides of the pit. The presentation also contained conventionally shot slides. These images were programmed to move through all quadrants of the screen area, thus presenting oriented images to all viewers.

About 300 slides—with 35 mm square apertures—were used in the presentation.

The format for the screen in the hood of the structure was relatively simple. The screen was divided into quarters, with two projectors (a dissolve pair) projecting into each section. The four additional projectors provided overlapping images across the centerlines of the screen. Full-screen images were created by using a combination of eight projectors. The slides used to create the full-screen images were mounted with two types of specially created "super seamless" masks.

The presentation was controlled by four QD3 units, manufactured by Audio Visual Laboratories, Inc. Each bank of three projectors was powered by an AVL Mark VII power module. The sound track for the presentation was played on a MacKenzie continuous-loop tape recorder. This equipment was placed outside **The Image Pit,** housed behind a smoke-colored Plexiglas door. This arrangement allowed viewers to see the programming equipment in operation, while still providing equipment security.

The presentation was programmed on the AVL Eagle.

The programming and audio equipment for the presentation were contained in an exposed console adjacent to the **Pit.**

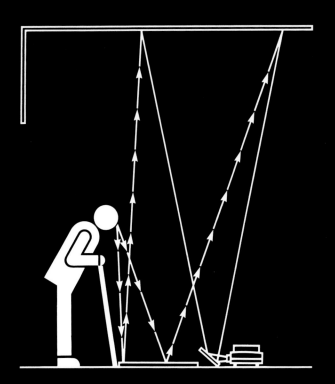

THE BASIC PLANNING DECISIONS— THE MASTER PLAN

Visual Style: What You See Is What You Judge

Almost everyone agrees that clothes don't make the man or woman. Yet almost everyone attaches a great deal of importance—consciously or subconsciously—to the way others appear. And this process of judging on the basis of appearance goes beyond clothes. We form opinions of people based on the way they walk, the gestures they use, the cars they drive, and the size and location of the houses they live in. In short, we judge people to a large extent on the basis of what we **see.**

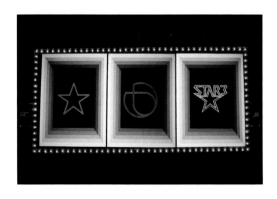

A presentation's visual style is more than "window dressing"; it's the emotional, subjective emphasis given to a presentation's message. (From Clear Light Productions, Inc. presentation at NAVA 1979, produced by Marsac Productions, Inc.)

This process of judging on the basis of appearances also extends to multi-image presentations. No matter how important the message you seek to convey, people are also going to judge your work on the basis of what they **see** on the screen. That's human nature: We tend to mentally evaluate the message and emotionally react to the medium. In other words, we are more objective about the former, more subjective about the latter. Check your own experiences to see that this is true. Think of the times you've left an audiovisual presentation saying, "I accept (or reject) the message and/but I thought the show was terrific (terrible)." The first appraisal is rational and objective; the second is more emotional and subjective.

So how your show "looks" becomes just as important as what it "says." Because of this, the matter of visual style is critical to a presentation's success. Visual style is the "look" of a presentation; it's the combination of screen format and projected images that audiences **see,** as opposed to program content, which they **absorb.**

How And Why Visual Style "Works"

Visual style enhances a presentation's message in three ways:

1. Style carries a large part of the emotional content of your message. It engages an audience's feelings. The message of your presentation, which should reflect your communications objective, is rational. It says "Buy this product" or "Give to this cause" or "Appreciate the meaning of this historical event." You may think—as many people do—that a straightforward, rational presentation of this message is sufficient to influence an audience. Unfortunately, that's not true.

Psychologists tell us that while people like to think they act on the basis of facts, they usually act because of their feelings. They may process facts; they may weigh options objectively. But when it comes time to decide or act, they do so because their rational, objective conclusions "feel" right to them. Emotionally, they're **comfortable** with their decision or action.

So, the "style" of your presentation accomplishes an important task. It makes your audience "feel" comfortable with — and accepting of — facts and arguments you're presenting.

2. Visual style intensifies your message. Style adds "punch" to facts, figures, and arguments. For example, suppose your presentation seeks to persuade viewers that your company's new line of products is styled to appeal to 18- to 34-year-olds. Your words may claim "youthful styling," "sleek lines for people on the go," "as bright and carefree as the people who use them." But by themselves, the words are unconvincing; they state an opinion. But show these products in a presentation with a youth-oriented visual style — contemporary, casual, and colorful — and you've intensified your message. You've said, "See for yourself, it's true."

3. Visual style sets the tone of a message. If you took courses in composition and rhetoric, you'll recall that tone is the manner or emotional stance an author uses to express his or her meaning. One author, upset by some aspect of society, will use a tone of anger to express strongly held feelings. Another may use a rational tone — a factual and objective analysis — to express the same opinion. A third author may adopt a tone of humor to satirize or ridicule what he or she sees. Each author is writing about the same subject — maybe even using the same facts — but each is stating an opinion in a different way. The difference reflects each author's hopes that the "overtones" will help drive the message home.

Tone, then, is the "vibrations" you add to your message to gain impact. To create those vibrations, you must first give thought to exactly what tone you want to set — a decision often overlooked by producers. Then you must determine how you'll use visual style to set this tone. If you're selling a product, for example, you'll probably want to set a tone of credibility. To do this, you could use a visual style that stresses the use of candid photography. If you're trying to raise funds for some social cause, you might want to establish a tone of concerned anger. Your visual style might be more artistic, using camera and film like palette and canvas to paint a photographic picture of suffering and desperation. When you set the tone of your presentation in this way, your rational message stands out, demanding to be heard.

A highly sophisticated visual style such as this, which combines creative photography with an artistic background, can draw audiences into greater emotional involvement with a presentation. (From Audio Visual Laboratories, Inc. presentation at NAVA 1979, produced by Synergetic Media.)

The Types Of Visual Style

The visual style of a multi-image presentation grows out of two factors: the artistic sensibilities of the photographer or illustrator and the underlying theme of the presentation. The first of these factors is difficult to classify. Each photographer and illustrator is unique and sees both the world and the technical requirements of photographic art in an individual way. Ask three photographers or illustrators to interpret the same subject and you'll wind up with three different viewpoints—and three different styles.

But photographers and illustrators usually aren't that free when working on a multi-image presentation. Individual styles are harnessed—or, more appropriately, they're **directed**—to give the proper visual emphasis to a presentation's theme. And these styles of visual emphasis can be classified.

1. The "straightforward" style. This approach doesn't mean "unimaginative." It means unadorned, a style described best by a phrase from architecture and design: Less is more. This type of photography eliminates the unessential, thereby placing emphasis on the true subject. Because the photographer has total control over the elements to be included or excluded from the picture, this approach could also be called a "staged" or "art directed" style.

A straightforward product shot, such as the one shown at the right, is a good example of this style, which is used most commonly to convey visual information—size, location, relationship of parts, color, appearance, and so on—in a clean, unambiguous way.

2. The photojournalistic style. This style emphasizes the candid presentation of events, places, and people. It's the style made popular by **Life** magazine.

This style works best when the communications objective of a presentation is "to show life as it is." An excellent example of this type of visual style is found in the presentation, **Around the Clock: 24 Hours in Blue,** a multi-image look at the operations of the New York City Police Department, produced by the Anders Group, Inc. of Staten Island, N.Y. The photographs for this presentation—one of which is shown at the right—were taken with a motor-driven 35 mm still camera, to capture life **as it is** for New York City police.

3. The artistic style. This style places emphasis on the classic concerns of all visual art—color, composition, perspective, etc. The primary intent of this style is not to convey information or to portray reality, but often to recreate life as art. This visual style dominates presentations focusing on the people, places, and customs of foreign lands, as illustrated by the photograph at the right.

An example of a "straightforward" visual style.

A shot from **Around the Clock: 24 Hours in Blue** illustrates the use of a photojournalistic visual style to capture—and depict—realism. (Photo: Anders Group, Inc.)

The artistic visual style emphasizes color, composition and perspective.

4. The abstract style. This visual style attempts to create symbols out of life, as the photograph below illustrates. To create this sort of visual style, photographers use special films and lenses, as well as laboratory techniques such as solarization and posterization. This style is effective when a presentation's message is broad, general, or abstract, a discussion of principles or ideas that might be weakened if tied to very specific visual images.

An abstract visual style uses nonrepresentational designs and symbols to convey meaning.

5. The graphic style. This visual style uses graphics and symbols to convey a message. At its simplest, this style involves the use of slides containing key words or phrases used to reinforce a spoken message. At its most sophisticated, the graphic style compresses the essence of a sequence into symbols that can be understood by international audiences. The slide at the right exemplifies a symbol with universal meaning.

Producers use the graphic visual style for much the same reasons they use the abstract photographic style: to avoid weakening or misrepresenting a broad discussion through the use of very specific visuals.

6. The illustrated style. Artwork and cartoons form the major elements of this stylistic approach. The illustrations can be as elaborate as the one shown at the right. They can also be as simple as stick figures. The illustrations can be original artwork, commissioned by the producer, or they can be photographs of paintings and drawings, copied—with permission—from art books, gallery collections, and historical texts.

Illustrations of this type serve as substitutes for the reality they depict. Producers use this style when the visual subject matter of a presentation isn't available, or when excessive detail within a subject might confuse understanding.

An example of the first application is a presentation produced for use at an historical site. A photographic treatment of the site and its importance might look too "staged"; it could detract from the "historical" component of the message. So illustrations—either original art or photographic copies of period artwork—could be used to illustrate the presentation.

An example of the second application is the use of illustrations dealing with such complex visual subjects as human anatomy or the electronic components of a computer. In the first of these cases, the subject matter may prove too detailed — or too explicit — to photograph, so artwork is used to illustrate and highlight relevant details. In the second case, the subject matter — memory chips — is too minute and complex to photograph, or to understand through photographs alone, so illustrations could be used to focus on key elements of the message.

Dividing these visual styles into six categories gives the impression that they are used exclusively; if you begin with one style, you must stick with it to the end. That's not quite true. While one visual style should dominate a presentation, you can use, where appropriate, elements of other styles. But this mixing of styles must be done artistically. You normally wouldn't mix an artistic style of photography with cartoon art. Nor in most cases would you want to mix a photojournalistic style with an abstract style of photography — unless, of course, your purpose is to move from an examination of the specific to a discussion of the general.

So when you mix visual styles, you should have a purpose — a definite reason for altering your visual "look." If you don't, if you mix different styles of photography with a variety of illustrations, cartoons, and graphic symbols, you're going to end up with a patchwork visual style. This is one reason why presentations using slides from dozens of sources — some old, some new, some original, some dupes — are rarely effective. They are stylistically inconsistent — analogous to a man dressed in top hat, tails, and sweatpants.

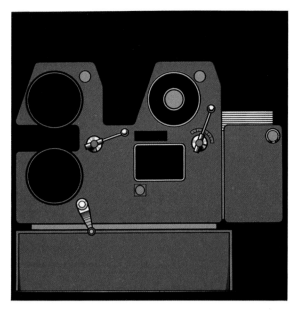

The graphic visual style is created from easy-to-understand designs and symbols.

The illustrated style relies on artwork and cartoons to carry visual information.

DETERMINING THE VISUAL STYLE OF YOUR PRESENTATION

You should approach the task of determining a visual style in much the same way an art director approaches the task of designing a magazine. You want to interpret your presentation's message visually, so you begin by asking yourself: What am I trying to say in this presentation?

Once you know what you want to say, you must determine the best way to depict it. Your first decision might be to decide between photography or some form of illustration. Once you make that decision, you have to decide what specific type of photography or illustration will serve your purposes best. The six descriptions above should help you match style to message.

Finally, you must choose the method you'll use to create your visual style. You can choose from three basic creative approaches: straight photography, photographic effects, and type and art.

Straight photography. This includes the production of slides—whether of the straightforward, photojournalistic, or artistic visual styles—using conventional films, cameras, and lenses. If this is the photographic technique you plan to use to produce your slides, no special planning is required. The right equipment and talent are all that you need.

Photographic effects. When you choose to modify a conventional photographic image—either in shooting or processing—you're using photographic effects. These effects can be created in a number of ways.

• **Special films.** By using special films, such as high-contrast, infrared, or X-ray films, a photographer can create unusual abstract effects. All of these films are highly selective in what they "see."

High-contrast film reproduces a scene in two tones—black and white; minor details are eliminated, and the effect is that of an abstract pen-and-ink sketch.

Infrared film is sensitive to the presence of heat; conventional details in a scene are thus ignored and an abstract impression of the subject is reproduced.

X-ray film is sensitive to the relative physical density of objects in a scene; the resulting radiographs combine the concrete and the abstract, the internal structure of a subject without its surface detail.

• **Special lenses.** Wide-angle lenses, telephoto lenses, and zoom lenses can be used by photographers to create effects that fall outside the boundary of normal vision.

Wide-angle lenses—including the so-called "fish-eye" lens—cover wide fields of view. Wide-angle lenses enable a photographer to show more of a scene than would be possible with normal vision. Used in the artistic style, such a lens also allows the photographer to distort images to create surrealistic effects.

An example of a straight photographic technique.

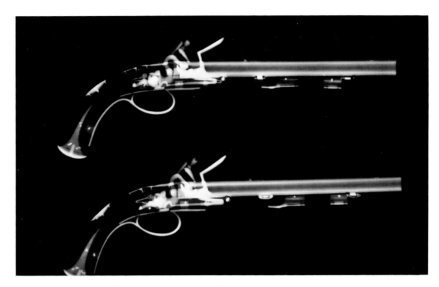

An image taken with X-ray film emphasizes "below the surface" details.

A photo derivation effect created by filtering high-contrast images.

An aerial photograph of a suburban neighborhood taken with infrared film is used to create a surrealistic image.

Telephoto lenses can be used to create foreshortened images, with foreground and background in full focus and appearing to be closer than they normally are in life. The photograph at the left shows the artistic use of a telephoto lens to compress a city street into a succession of small storefronts and signs.

Zoom lenses, besides allowing the photographer to quickly frame a subject, can also be used very creatively to produce exciting visuals. One of the most impressive techniques is to zoom the lens during a long exposure, to add a look of motion to the subject.

A telephoto lens compresses elements of a scene into a single plane.

Image of city skyline at night explodes with light and motion when focal length of zoom lens is changed during exposure.

• Special-effects filters. Although some purists avoid the use of special-effects filters, there are occasions when they can be used to advantage — to add life to an otherwise dull subject or to make the specific become generic. The wide assortment of filters on the market today makes it impossible to show or even describe them all within the confines of this book, but some of the more common and dramatic ones are illustrated on these pages.

Spot filters give your images a soft, almost ethereal look. They can also eliminate extraneous subjects within your field of vision.

Diffraction-grating filters create slivers of shimmering light from any point of undiffused light in the image area. They come in several configurations that can be used separately or in combination. More recent types also break the light down into the colors of the spectrum.

Multi-image filters also come in a wide variety. You can multiply the subject with a series of parallel images or with a number of images encircling a central one.

Color filters, available in many different colors, can add excitement to what would otherwise be dull photographs.

• "Sandwiches" and "burn-ins." The sandwich slide combines two or more transparencies in a single mount. This type of slide is typically used in two ways:

One application allows a producer to create superimposed images for projection more easily, and with greater control, than trying to create the same effect through double exposure in the camera. Separate images, both slightly overexposed, are mounted in a single mount. When projected, they form one image.

The second application allows a producer to incorporate type or graphics with photographs. With this technique, the photographer shoots two slides. The first is of the actual subject, shot in such a way as to leave a suitably large light-colored area as background for the type. The second slide is of white type or graphics (on a black background) that is copied on Kodalith ortho film, type 3. (The white image on black background is usually a photostat of black type on a white background.) When processed, the high-contrast negative will appear as a black image on a clear background. To create the "sandwich," the producer joins the photographic transparency and the negative in the mount, with the type or graphic over the light area of the slide.

"Burn-ins" are similar in effect to an image-and-type sandwich slide, but their appearance and production are different. There are two methods of creating a burn-in. In the most common method, a slide is shot of the subject (on reversal film), preferably leaving one area of the image slightly darker than the rest. Then a high-contrast slide is made of black type or graphics against a white background. The first image is then duped and double-exposed with the slide, producing a slide with white type or graphics "burned in" on the image. (By dyeing the high-contrast slide or using filters over the lens during the second exposure, colored type or graphics can be produced. When attempting this, however, it's important that the color of the type or graphics be light enough to burn in on the primary image.)

The second method of producing a burn-in is basically the same, but is accomplished by double-exposing the reversal film during original shooting. Because this requires shooting the second exposure immediately after shooting the original subject, you'll have to prepare the high-contrast slide first. This is practical only under studio conditions.

Spot filters soften images.

Diffraction-grating filters emphasize subject with slivers of light.

Multi-image filters create an array of images.

Color filters enable photographer to "paint" images to create surreal effects.

Two images "sandwiched" to create surreal photographic effect.

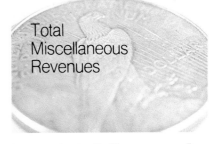

Type "sandwiched" with a photographic background.

Type burned into a graphic background.

Professional photographic and darkroom techniques are explained in Kodak's **Here's How** books available from your local photo dealer.

Masks and mounts are an easy way to add variety to a visual style.

• **Laboratory techniques.** Many producers use special processing techniques such as "solarization" and "posterization" or "photo derivations" to create special effects. Detailed instruction for producing these effects can be found in numerous photographic publications such as the "Here's How" Series from Kodak.

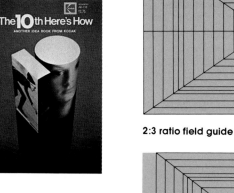

2:3 ratio field guide

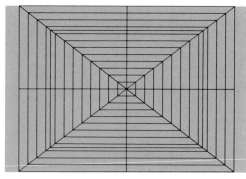

3:4 ratio field guide

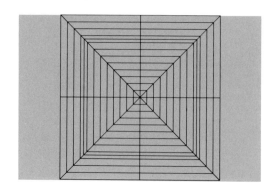

1:1 ratio field guide

Almost all of these photographic effects require extra planning on your part. Special films often require special cameras and equipment. In the case of X-ray photography, you also need a trained technician to assist you. If you're using special lenses, they may have to be acquired. When using sandwich slides or burn-ins, you must plan your transparencies in terms of exposure and composition. And if you're planning to use laboratory techniques, you must locate a lab that can process your work. If you ignore these planning steps now, you may not be able to achieve the effects you want later.

Art and Type. Art slides—including illustrations, graphics, cartoons, and copied material—present two major planning problems. First, you must make certain your illustrator creates the work in the same aspect ratio as your slides. If you're shooting 35 mm slides and your illustrator composes the work in a square area, you're going to end up with slides with wide borders on each side or cropped top and bottom. On the left you'll find illustrations of artwork templates—for the common slide aspect ratios—with suggested dimensions for work areas.

Second, to make it easier to photograph the artwork, have your illustrator mount the work on pin-registered boards. This is especially important if any of your sequences will be created using overlays.

A more detailed discussion of this subject will be found in the following section on production, along with the associated subject of type legibility on slide art.

Mounts and Masks. These aren't creative methods per se, although they can contribute significantly to your visual style by creating unusual "frames" for your images.

You can create these frames using commercially available mounts and masks. Several suppliers are listed in the appendix. Or, you can create—through sandwiching techniques—your own frames, using masks made from Kodalith high-contrast film. This approach will be described in detail in the Production section.

Planning "With Style"

Let's say you've been asked to produce a show for a university fund-raising drive. The communications objective: to persuade people to donate the equivalent of one day's salary to the school. The theme of the presentation: Give us one day for the four years we gave you. In planning the visual style for this presentation you would:

1. Ask yourself what you're trying to say. In this example, the answer is, "The education University A provided has contributed to your present success; now we ask you to help us help others."

2. Determine how to illustrate this message most effectively. Since this is an appeal for money to support university activities, you might choose a photojournalistic style, hoping to illustrate factually—and support emotionally—the idea that the university wisely spends the money it receives. You might show students in classes, laboratories, and libraries; you might also show faculty members preparing for classes, leading class discussions, or counseling students. In short, you would show "the way life is" within the university, with the emphasis on the people and activities that benefit from financial contributions. To substantiate your plea for funds, you also might use a graphic style to present statistics highlighting the costs involved in educating students at University A, and the rising costs of education in general.

3. Decide on the creative methods you'll use to build your visual style. If you're using a photojournalistic style, very little additional planning may be required. Finding a photographer with strengths in candid shooting should be sufficient. Planning would be required, however, if graphic slides are to be used. You would have to choose the creative method: hand-drawn lettering, straight type, or type-and-transparency sandwiches or burn-ins. Considering the intent and sophistication of the hypothetical presentation, you would probably choose a type-and-transparency treatment.

With these steps taken, the overall visual style of the presentation would be set.

TWO APPROACHES TO ORGANIZING YOUR VISUAL STYLE

You can organize the visual style of your show in one of two ways. In the first, you work from an actual script, noting in the margins the types of slides you'll want to use. These notations are referred to as "call-outs."

Call-outs don't list each and every slide you'll use; rather they indicate the types of slides you'll want for specific sequences. For example, if the script for the presentation on University A's fund drive contained a section on the growth of the school's engineering department, you might note call-outs such as: "Show students in classroom with blackboards covered with equations; also show close-ups of intent young faces; extreme close-up of student notebook (with sufficient light or dark areas for graphic sandwiching or burn-ins depicting department growth)." With these instructions, you or your photographer can shoot the slides that will best convey the content of that portion of the show.

In the second approach, you create a storyboard (or planning board) to gain greater control over the planning of your visual style. A storyboard is a series of illustrations showing the exact composition and camera angle for each slide in a presentation. (See example at right.)

The producer gives the completed storyboard to a photographer. If the storyboard is detailed enough, the photographer will need no other instructions. He or she will reproduce the images indicated in the illustrations.

A potential drawback of this approach is cost. If you're planning a presentation of 30 minutes or more, you could spend a considerable amount of time and money creating a storyboard containing 800 or more illustrations.

Another possible drawback of storyboards is that they may limit the creativity and spontaneity of the photographer. If a storyboard is conceived and used as artistic "law," a photographer has little creative license. "Opportunity" shots—those unpredictable situations that occur during the course of shooting—will be bypassed, thus depriving the producer of slides that may better express the intended meaning of a sequence.

Additionally, duplicating the composition of a storyboarded image often leads to "staged" shots that seldom have the credibility of true candid shots.

Of course, sometimes the controlled planning imposed by a tight storyboard is desirable. When a presentation will consist almost entirely of specific subject setups, the producer won't want the photographer "interpreting" the assignment.

The copy in the left column consists of "call-outs"— verbal descriptions of a presentation's visual development.

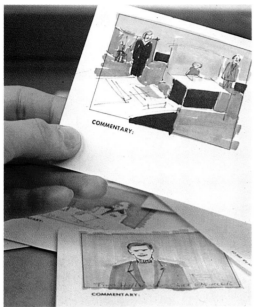

How Many Slides Will You Need?

Not only does a visual style affect the "look" of a show, it also affects the number of slides you must shoot to produce the show. Less controlled styles, such as the photojournalistic style, usually require more shooting than controlled styles, such as the artistic or graphic styles.

Before you begin actual production, you should estimate the number of slides you'll need for the final presentation. This figure will help plan for the amount of film, the length of time, and perhaps, the number of photographers you'll need to complete the assignment.

A simple formula will help you estimate the number of slides you'll need:

1. Determine the length of the show in seconds.

2. Multiply that figure by the number of screen areas being used. (For example, if you will have a bank of three projectors aimed at the left side of the screen, a bank of three projectors aimed at the right side of the screen and two banks of three projectors aimed at the center overlapping area, the figure would still only be 3, because there are only three screen areas.)

3. Divide this figure by the average length of time—in seconds—that an image will be on the screen. (A figure about 4 to 6 seconds would be average for a moderately paced show.)

The result is the approximate number of slides you'll **need** for your presentation. To determine the number of slides to **shoot**, you have to determine the probable shooting ratio during production. (Shooting ratio expresses the relationship between the number of slides shot and the number actually used. If your show is to be based on a photojournalistic style, you may reach shooting ratios of 15 to 20 slides shot for every one used. If your show is to be based on artwork, your shooting ratio may be as low as 4:1.)

4. Take the number of slides you need and multiply that figure by the first figure in your shooting ratio. The answer is the probable number of slides you'll have to shoot to arrive at the number you need.

If the figure you arrive at seems unreasonably high, don't cut it back without good reason. In planning for a multi-image presentation, it's better to overestimate the number of slides you'll need than to underestimate. It's always easier to discard slides you don't use than to go out and reshoot a sequence that needs additional visuals.

PLANNING FOR THE PRESENTATION SITE

At the beginning of a production, most of your attention is focused on the presentation elements—the script, photography, sound track, art, screens, and programming equipment that form the heart of the presentation. Because of this, there's a tendency to overlook several planning factors related to the actual presentation—the dimensions and seating arrangements, the acoustics, and the power supplies at the presentation site.

In the unit on planning the visual format, you examined the first of these factors—room dimensions and seating. You may have already planned the seating and the placement of screen and projection booth. Now, to complete your site preparation, you must plan for the sound portion of your program.

Acoustics. Gone are the days when audiences accepted scratchy sound tracks and muffled voices in audiovisual presentations. Today people are used to hearing studio-quality sound in their own living rooms, so naturally they want to hear the same quality when they sit down at a multi-image presentation. The producer who overlooks this requirement is diluting the impact of his or her program.

Despite the acoustical sophistication of audiences, sound remains a neglected aspect of multi-image planning. It's still taken for granted by many producers who rely on a high-fidelity speaker or two near the screens for the audio portion of their shows.

Planning for the audio portion of an effective presentation requires considerably more preparation. It involves studying the acoustical properties of the presentation site, then selecting and placing loudspeakers to create a balanced sound pattern. It also involves selecting and linking the sound components (tape recorder, motion picture projector, microphones), the amplifier, and the loudspeakers.

This planning isn't a major problem if the presentation site is acoustically balanced—neither absorbing nor echoing sound. But most of the sites you'll work in—conference rooms, hotel meeting rooms and exhibit halls—lack acoustical balance, so you're going to have to compensate for any sound-deadening or sound-reflecting properties.

Unless you have experience in planning sound systems, you'll need the help of an audio specialist. If such a specialist works on your staff, you have no additional problems. Just give the specialist your requirements and let him or her solve the problem of power, fidelity and sound balance.

If you don't have an on-staff specialist, you should be able to find one in the Yellow Pages of your telephone book. Look under the listing for Sound Systems and Equipment or Public Address Systems. In some larger cities, you'll even be able to find listings for AV Production and Presentation Services. The companies listed under these headings should be able to provide the audio specialists and equipment you'll need to create an acceptable sound system for your presentation.

These sound system companies are especially helpful if your presentation is to be shown at out-of-town locations. They can scout your presentation sites and recommend equipment. When the day of the presentation comes, they can also install and test all equipment.

You may be tempted to put off this planning until later in your production cycle, when the presentation is beginning to take shape. You may feel you have enough to worry about now and don't want to get involved in additional details, especially those that seem unimportant at this early date. Admittedly, it's a natural temptation, but you should resist it.

Sound should never be a last-minute consideration. It should be a part of your planning from the time production requirements are set until the day the system itself is set up at the site. Only in this way can you be certain that the quality of your sound will equal the quality of your images, that people in the last row of your audience will hear just as clearly as the people in the front, that the sound track your audience hears has the same crispness as the sound track you first recorded.

A specialist in audio systems is usually necessary to help plan sound requirements for large presentation areas.

Power requirements. This is another area for a specialist. There must be sufficient electrical power available to operate **all** the equipment needed to put on your presentation. And this power must be "uninterrupted," a term meaning the source can't be disconnected, tapped, or short circuited by someone else. It is power dedicated solely to the presentation equipment. (Also remember that starting up a projector motor usually requires more current than running it. Allow a safety factor in your calculations.)

The only way to assure you have this power is to work with the facilities management staff of the building in which you'll put on your presentation. You must let these people know how much power you'll need (in total amperage), when you'll need it, and for how long (and don't forget you'll need time for setup and rehearsals).

You can use the chart at the right to determine your power needs.

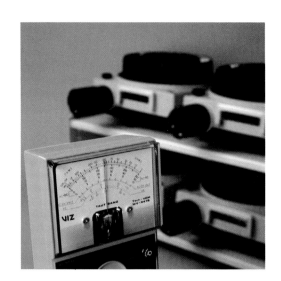

Manufacturers of presentation equipment list electrical power requirements on their products.

Equipment	Amperes	×	Number	=	Power Requirement
Slide projectors	_____		_____		_____
Motion picture projectors	_____		_____		_____
Dissolve modules	_____		_____		_____
Programmer(s)	_____		_____		_____
Tape deck	_____		_____		_____
Sound system (amplifiers, mixers, equalizers, loudspeakers)					
Spotlights	_____		_____		_____
_____	_____		_____		_____
_____	_____		_____		_____
_____					_____
	Total Power Requirement				_____

Tips For Planning Power Requirements

1. Some companies list the power requirements for their equipment in watts rather than amperes. If this is the case with your equipment, you can find your amperage requirements by using a simple formula: **Watts divided by line voltage equals amperes.** For example, if you're using a projector marked 400 watts with a 120-volt power supply, perform the following calculation:

$$\frac{400 \textbf{ watts}}{120 \textbf{ volts}} = 3.33 \textbf{ amperes.}$$

2. When calculating power requirements for dissolve units, remember that the "watts rating" on the unit is for maximum load—the amount of power the unit can handle for its own needs **plus** the power needed to drive its projectors. For this reason, it's better to figure your power requirements starting with your projectors; then include an allowance for the dissolve unit's requirements. For example, if you're connecting two projectors (with 300-watt lamps) to a dissolve unit, calculate power requirements like this: **350 watts for each projector (300-watt lamp plus 50 watts for the motor) plus about 100 watts for the dissolve unit itself.** Using the formula you would calculate:

$$\frac{350 + 350 + 100 \textbf{ watts}}{120 \textbf{ volts}} = \begin{array}{l} \textbf{6.66 amperes for} \\ \textbf{dissolve unit and} \\ \textbf{two projectors.} \end{array}$$

3. Your planning for electrical power must also consider the problem of improper "phasing"—a condition that exists when interconnected units of equipment, especially dissolve units and projectors, are plugged into outlets with different polarities. When this happens, the non-synchronous electrical phases may cause equipment malfunction (i.e., unwanted slide projector cycling). The simplest remedy for this problem is to plug **all interconnected** equipment into a common power source.

MOVING FROM PLAN TO SCHEDULE

Organizing Your Efforts

The principal lesson of this chapter can be summarized in a single sentence: **Plan your work and work your plan.** It's a simple concept, but one many producers find hard to put into practice. Their difficulties are usually three: They don't know where to start planning, they don't know what elements to include in their plan, and they don't know how to organize their planning decisions in a format that becomes a useful production tool.

Thorough planning leads to a smoother production.

One solution to their problems lies in a planning method called Critical Path Management (CPM). CPM helps a planner organize and schedule the people, resources, time, and money needed to produce a multi-image presentation. Equally important, CPM provides a method of **graphically displaying** this organization and schedule—the CPM chart. This chart becomes a guide to—and a constant reminder of—the work that must be performed to produce a show.

CPM offers the planner three advantages:

- CPM charts are easy to prepare. Major activities are divided into manageable tasks that are plotted in a logical and chronological sequence from start to finish. Sequential tasks are plotted horizontally, from left to right, to show the progressive development of the various phases of a project. Each of these "paths" is also given a priority.

- CPM charts are easy to understand, so they become an effective way to communicate decisions and priorities to others.

- Most important of all, CPM charts organize efforts in a way that produces results. The planner/producer and his or her staff know the number, priority and order of the tasks that must be accomplished. There's no "wheel spinning," no time-consuming delays, no overlapping of effort, because a CPM chart isolates each day's (and each week's and each month's) crucial activities.

In short, CPM methods help you plan your work. Later, in Section Three (Production), we'll show you how to work your plan.

Planning Equals Preparation

When you sit down to plan, you're following the advice summarized in the Boy Scout motto: Be Prepared. That's what planning is—a method of preparation. It enables you to anticipate problems. And when you anticipate problems, you have an opportunity to avoid them or to solve them before they become major obstacles to production.

The key to successful planning is to separate production activities into those that can be performed simultaneously and those that must be performed sequentially. To understand how this process works, let's look at the activities involved in a typical multi-image production.

You have to research your subject and write a script. You have to shoot slides and, if your presentation requires it, shoot motion picture footage. Art may have to be prepared and photographed. Slides and film must be edited and a sound track must be narrated and mixed. Screens and projection equipment must be selected—perhaps even ordered. The show must be programmed and then rehearsed. Arrangements for seating, audio, and projection at the presentation site must be made. Screens and projection equipment must be shipped to the site.

These are major activities. Many other minor activities and details also must be planned for. It adds up to a lot of work, and if you tried to perform all these tasks separately and sequentially, you would be months producing even the simplest multi-image production.

Of course, you wouldn't tackle a project this way. Even if you've never produced a multi-image presentation, it should be obvious that a complex project has to be launched on several fronts simultaneously if the work is to be completed in a reasonable amount of time. But unless you've had experience in scheduling, what you probably don't know is that there's a definite sequence to follow in launching a multi-phased project. The natural impulse is to take "first things first." That advice may be appropriate if you are organizing a simple project. But with multi-phased projects, especially a multi-image production, the rule to follow is **"longest things first."**

The "longest things first" rule forms the heart of planning techniques such as Critical Path Management and PERT (Program Evaluation and Review Techniques). PERT is the more sophisticated of the two techniques and usually relies on probability formulas and computerized monitoring systems for its effectiveness. Critical Path Management is a simpler planning system based on PERT techniques. The most important step in PERT or Critical Path Management is to isolate the chain of events that will take the longest to complete. This chain runs from event Number 1, which kicks off a project, to the final event, which brings the project to a close. This chain of events is called the "critical path." (A sample critical path chart appears on page 239.)

The critical path receives the planner's highest priority. Each of the events that form the path must be completed on time if the project itself is to be completed on time. Any delays in the critical path signal trouble for the entire project.

The critical path and the other parallel "paths" on the chart link sequential activities, those that must be performed in the order indicated. Each activity builds on the results of the previous activity and contributes to the completion of the subsequent activity. The effect is cumulative, much like the building of a house. First the foundation must be built, then the first floor, then the second, finally the roof.

The chart's vertical orientation helps pinpoint those activities that can be undertaken simultaneously. For example, early in the production cycle, while research and scripting are under way, a producer can also begin ordering the equipment he or she needs and planning the seating and staging arrangements for the presentation site. None of these activities needs the results of the other two for the work to be completed. So they can be undertaken at the same time.

By strategically planning sequential and simultaneous activities in this way, a planner can schedule time and people most effectively and most efficiently. Tasks are started and completed in a definite order, and delays and idle time are kept to a minimum.

Understanding Critical Path Planning

Perhaps the best way to understand how critical path planning works is to study the hypothetical plan on page 239. This plan includes just about every activity you might encounter in the production of a typical presentation—scripting, artwork, slide production, film production, laboratory services, the ordering of equipment, etc. (For the most part, you won't perform every activity we've listed during a single production. But sooner or later you will have to plan for each activity on the chart. Because of this, you should know how to fit all of them into your scheduling.)

The first item you should examine is the wide blue path running horizontally through the middle of the chart. That's the critical path. This path includes the activities necessary to produce a motion picture sequence: shooting, editing, conforming, approval of answer print, and the printing of release prints. This sequence of activities is "critical" because it will consume more time than any other parallel sequence in your production cycle.

Next study the paths indicated by the slightly narrower, lighter lines. These indicate what could be called "secondary paths." Included in these paths are such activities as the production of photo derivations (posterization and solarization), the use of animation or copy-stand photography, the duping of slides, and the ordering of production and presentation equipment. These activities are given a secondary priority because they too require considerable time, but not as much as those included in the critical path.

In addition to these primary and secondary paths, you'll see a number of lesser (third-level) paths. (They're of lesser importance only in terms of scheduling priorities, not in terms of eventual contribution.) These paths indicate activities whose scheduling is determined by the requirements of the critical path. Work on these activities is scheduled for periods when critical path activity slackens.

Planning for the selection and ordering of equipment is a good example of these third-level paths. As you can see from the chart, the producer/planner has greater leeway in scheduling activities along this path. But rather than to randomly assign time to these tasks, the planner looks for periods where critical path activity is reduced. He or she then schedules transportation arrangements into these "free" periods.

You should notice one other characteristic of a critical path chart. All activities are listed in the present tense, as tasks to be completed. Listing activities in the present tense — in effect, as commands to you and your production team — acts as a motivating trigger. You know what you have to do: **set** communications objectives; **order** projection equipment; **write** script; **edit** slides; **program** presentation. The imperative tone of these statements will remind you and your staff that constant effort and continuous progress are necessary to meet your many deadlines.

One final point: We haven't listed dates or in any other way indicated elapsed time between activities. It would be impossible to do so; each production presents its own requirements, and those requirements determine schedules and deadlines. You, however, must indicate completion dates on your CPM chart or on a schedule drawn up from the chart. If marking deadlines on the chart, we suggest the dates be placed inside the circles marking the termination of activities. There they can't be missed — or avoided or ignored.

What The Chart Means To You

You've seen, in broad and general terms, how critical path planning works. Now let's try to get a bit more specific, examining the planning implications of these various paths.

1. If your presentation is to include film sequences, production of these sequences must begin as soon as possible. If you have sufficient time, you can delay shooting until after script approval. But if time is critical, you must try to accelerate this activity.

One way to gain time is to begin shooting before script approval — if you can. This is usually possible when the film sequences will set the mood and visual background for a presentation, but will not carry vital visual information related to the presentation's theme.

If you must wait for script approval, use the time to plan the logistics for and the action of your shooting sessions. Don't wait to begin preparations until after the script is approved. Be ready to go as soon as you get an okay.

Another approach, one that can prove useful when deadlines are short, is to ask clients to concentrate their initial efforts on reviewing and approving the film segments of a script. In this way, you're able to begin work on the more "critical" film sequences before the total script is approved.

2. If you plan to use photo derivations, motion picture animation, or copy-stand photography, begin these activities as soon as possible. They take longer to complete than conventional photography; so the sooner you start them, the sooner your slides will be available for editing and programming.

3. If you must order programming or projection equipment for the presentation, do so as soon as you've selected the units you'll need. Depending on the volume of incoming orders, manufacturing schedules, and inventory levels, manufacturers and dealers may need six to eight weeks **or more** to fill your order. If you haven't planned for a wait of this duration, you could wind up without the equipment you need to program and present your show.

4. If you plan on producing duplicate sets of slides for your show, allow yourself a reasonable amount of time for this activity. And remember, you also may have to allow for more than just lab time if you plan to manually remount the duplicate slides in any special type of mount.

97

Planning— From End To Beginning

The most effective way to plan is to work backwards. Determine when your presentation must be completed; then work back from there, figuring out when each of the various contributing phases must be completed. Keep working back, from major phases to subphases to minor activities, until you reach your starting point. Then, if the chart you've prepared looks logical and workable, you're ready to begin production.

Planning from end to beginning doesn't reduce the amount of work you have to perform, but it does make it simpler. For one thing, this back-to-front approach makes it easier to identify the activities and the people you need to complete each task. Working in the normal fashion, from beginning to end, often results in abbreviated planning. The planner may make assumptions along the way as he or she concentrates on the completion of major tasks. For example, a planner might include in the plan a notation such as "Equipment Delivered." The completion of this task, however, assumes the completion of several earlier tasks, such as who makes the transportation arrangements—and when.

Making such assumptions is easier when working beginning to end, but not when you're working the other way around. If, while working backwards, you make an entry for "Equipment Delivered," you're faced with several unavoidable questions: How will it be delivered? Who will transport it? And who will arrange for this transportation? With each question, you trace the sequence of activities back to its roots, to the decision to produce a desired result. By tracing the probable steps, you automatically touch all the "planning bases."

The end-to-beginning approach to planning also helps you fit all production activities into the time allowed. Again, if you're working beginning-to-end, it's easy to overshoot a deadline. At the beginning, you aren't quite sure where the critical mileposts in production must be scheduled. You're like an explorer trying to find his or her way through uncharted land. You have no landmarks to guide you. So you simply forge ahead, hoping to arrive at your destination.

When you work from end to beginning, however, you know where you have to be at specific dates, so you can adjust your earlier activities accordingly. If, as you plan backwards, you find the time allowed for shooting and processing slides is too tight, you can shorten the time allotted to scripting to gain yourself some breathing room.

What To Do If Time Is Short

There's almost no way to avoid it: Someday you'll find yourself planning for a presentation in which there's too much work and too little time. When that day comes, you're going to have to perform some surgery on your schedule.

This surgery can take two forms. You may have to add more "muscle" to particular production activities. Or, you may have to cut activities into parts that can be worked on outside the normal production sequence.

The first approach is easier—if you can afford it. What it entails is cramming more hours of work into every working day. And the only way you can do this is by hiring extra people or contracting with additional suppliers. The period of time needed for photography, for example, can be shortened by hiring additional photographers, rather than by trying to do it yourself. Or you can contract with a sound specialist to edit and mix your sound track while you begin sorting and editing slides. This approach is highly effective—it gets the work done without cutting corners. But it's also expensive; you have to have the money in your budget for it to be a feasible option.

If the money isn't there, you have to rely on the second approach. You have to take segments of activity from your critical path and begin them "ahead of schedule." One area where this approach is effective is in the programming of the presentation. Under normal working conditions, you would begin programming after a script is approved and the sound track for the presentation is completed. When time is short, however, you can begin programming individual segments separately and out of sequence. You might program areas of the script where no revisions have to be made, or areas such as the beginning and ending where there are no "words" to be concerned about. Once programmed, these segments would be stored until you were ready to complete the final programming. Then, with some revisions to ensure smooth transitions, these previously programmed segments would be worked into the final program.

Naturally, this sort of divide-and-conquer approach is tougher to complete than the more natural, beginning-to-end sequence of programming. It requires more from you in terms of planning and reassembling the final program, but it does save time.

A Secondary Benefit Of Planning

A well-planned production cycle helps you organize your time and efforts, placing emphasis where it is needed and when it is needed. It also serves another important role. A schedule helps you plan your need for funds. You can estimate, by looking at your schedule, how and when you'll be spending your budget. For example, if you look at the sample schedule, you'll see that during the initial phases of the project, money is needed for scriptwriting and for ordering equipment. Once the script is approved, greater sums of money are needed to pay for the film and slide shooting and the sound recording that will take place during this phase of production. Once this phase is over, the amount of money needed drops somewhat as the presentation goes into programming. In the final phases of production, larger sums will be needed for transportation, travel, and setup.

This sort of budgeting information is important if your organization issues budgeted funds in monthly "allowances." By reviewing your planning chart, you can estimate, month by month, the amount of money you're going to spend. If you're an independent producer, this sort of spending information is important in negotiating a contract with your client. Most multi-image production contracts, especially those for major shows, contain clauses stating payment will be made in three or four installments, each coinciding with a major milepost in production. The planning chart allows you to identify those mileposts where spending will be greatest, thus allowing you to arrange payments to cover these activities.

BRINGING YOUR DECISIONS TOGETHER

The Project Proposal

Making decisions concerning your multi-image presentation involves one set of problems; communicating these decisions to your client involves another. So far, all your efforts have been directed at evaluating and choosing the various production options you can use to create a presentation. You've pulled together decisions on:

- the communications objective
- the presentation requirements
- the budget you have to work with
- the time frame you must work within
- the media you'll use
- the visual format you'll create
- the visual style you'll employ
- the preparation of the presentation site

Decisions and recommendations are gathered in the project proposal.

Now you have to sell these decisions to your client—whether that person is someone in your organization or someone from an outside organization. You have to convince your client that your ideas on how to convey the presentation's message are workable—creatively, aesthetically, logically, psychologically, and financially.

And to sell these ideas, you have to "package" them. You just can't jot down your decisions on a piece of paper and give them to your client. Or worse yet, tell him or her, "Here's what I'd like to do."

In these casual forms, your ideas mean little; they're just thoughts, related to the project, but not organized to give an impression of what you want to accomplish and how. Thus you must package your decisions so they appeal to your client. And you must present them in a form that allows your client to fully understand your conception of the presentation.

You package your thinking in a document called the project proposal. **The project proposal is a detailed, written explanation of what you hope to accomplish in a multi-image presentation.** It communicates your intentions—and relates them to your client's objectives. If, for example, the client wants to prompt inquiries about a new product, the project proposal explains exactly how your plan will trigger those inquiries.

For you, then, the project proposal is a decision-clarifying and decision-unifying tool. It forces you to group and give direction to a series of separate planning decisions. For your client, the project proposal is a decision-making tool. It enables him or her to study your decisions, not as separate ideas, but as elements of—and contributors to—a unified whole.

A well-written project proposal should contain five distinct elements: a statement of objectives, a theme and premise, a content outline, a budget, and a tentative schedule.

1. A statement of objectives. This is simply a restatement of the communications objective developed during the initial planning for the presentation. It states the behavioral change — the reaction — the presentation is intended to create in an audience. This statement can be as brief as a single sentence. Or it can be far more elaborate, explaining the problems underlying the objective and the probable consequences a change in audience thinking or behavior will have on the solution of the problem. The formula for expanding the objective should state why the problem exists, what the objective is, and how the presentation will contribute to that goal.

For example, a producer for a company trying to increase orders for a new product should state the objective briefly: "To increase orders for Product A." In fact, if the company came to the producer and presented him or her with the problem and the objective, there would be no reason to elaborate any further.

If, on the other hand, the producer felt it necessary to sell **both** the presentation and the reasons for producing it — as he or she might in an organization undecided on an approach to a communications problem — he or she would more than likely expand the statement of objectives, establishing a more detailed context for the presentation. This could be accomplished as follows:

Background of Problem: Product A was introduced to the marketplace during a period of recession and did not receive the publicity or attention it needed to gain wider acceptance.

Communications Objective: The goal of this presentation is to overcome this initial communications failure and to create widespread consumer awareness, resulting in an increase in orders for Product A.

Rationale: The presentation will create enthusiasm for Product A among two key groups — dealers and major industrial buyers. They will see Product A as a time- and money-saver — an easy-to-use, easy-to-maintain product that's indispensable in the modern office. Once these two groups are aware of — and interested in — Product A's benefits, the company will launch a major advertising campaign building on this enthusiasm.

2. A theme and premise. The theme of a presentation is its developing purpose. It's a statement of what the presentation will **prove.** Naturally, it's an outgrowth of the communications objective.

For example, the objective of a presentation might be "to increase orders for Product A." A suitable theme for such an objective might be: "Because it requires three fewer steps to operate, Product A increases the efficiency of office workers by 35 percent." This is the statement that must be proved.

The premise of a multi-image presentation explains the creative approach that will be used to develop the theme. It is the flesh on the skeleton of the theme. In the example above, the theme of "efficiency" could be developed using a premise that showed side-by-side comparisons of Product A and Product B. Another premise might be the use of on-the-job interviews with office workers who would tell of increased efficiency when using Product A. A third premise might be a before-and-after examination of office output. Each premise proves the theme in a slightly different way. The best premise is the one that involves the audience the most. Usually it's a premise that relates the presentation's theme to the experience of an audience.

3. Content outline. This is exactly what it suggests: an outline of the points to be made during a presentation. If the theme states what is to be proved, the outline presents the elements of that proof and the sequence in which the elements will be presented.

In high school or college you may have been taught to create outlines using Roman numerals, upper- and lower-case letters, and varying degrees of indentation. If you're still comfortable outlining your ideas this way, by all means do so. But for most people, the formal outline is a barrier to creativity. It's too rigid, too confining; it becomes a pattern to complete rather than a useful form for generating—and explaining—creative thought.

So, create your content outline in any way that works best for you. This might be a simple listing of the points to be made: 1, 2, 3, 4, 5 . . . etc. Or, it might be an unnumbered series of paragraphs describing the material to be covered. Or maybe it's a format of your own creation. It doesn't matter as long as it describes the thematic development—the proof and reasoning—of the final presentation.

4. The budget. You may or may not include this element in your proposal. If the budget for a presentation has been set and your only responsibility is to spend the money wisely, then, quite literally, you have "nothing to say." If the budget for your presentation is still open (for example, if a general sum for visual programs or conference activities has been set aside, but a specific amount for your particular presentation hasn't been established), you must include cost figures for your client's approval. In some cases, the client will require no more than an overall cost figure. Usually, however, you will have to break down this total into specific production costs: scriptwriting, photography, programming, music, etc. If you're not certain which course to follow, ask your client about his or her preferences.

5. Tentative schedule. In this section you should note the dates on which specific production mileposts are to be reached—for example, script completed, photography completed, programming completed, and so on. Equally important are the dates on which **you** must receive approval of such key elements as the script and the visual format so you can continue production. This information should be taken from your planning chart.

Creating a project proposal from the above five blocks of information isn't as much a task for the imagination as it is for the intellect. Remember, when you write a proposal, you're trying to persuade a client to accept your decisions. You're trying to convince him or her that your ideas and your logic are sound. So avoid using flamboyant language in your writing style. Stick to your facts and your arguments, and present them in a clear, straightforward, and coherent manner. A project proposal is a business document, and it should read like one.

Sections from an actual project proposal have been reprinted on the following page to help you in planning and writing your proposal.

Proposal for:

XYZ CREDIT ACCOUNTING SERVICES

Objectives

XYZ Credit Accounting Services finds itself challenging a number of established companies in the field of credit card accounting services. It must, therefore, work all the harder to gain access to decision-makers in banks and credit unions if it hopes to sell its contract services. The proposed multi-image presentation will help gain that access. Its principal communication objective is:

To convince prospects to enter into a formal business presentation for XYZ credit accounting contract services.

Theme

The theme—or basic argument—of the presentation will be:

Because XYZ Credit Accounting Services is located within your marketplace, it can enable you to respond to client needs more efficiently and effectively than can an outside processing service.

This theme will be developed in a 15-minute, six-projector, multi-image presentation highlighting the operations, responsibilities and advantages of the XYZ Credit Accounting Services. The emphasis of this presentation will be on the effectiveness and efficiencies of XYZ operations, leading to the action step: "Allow XYZ to develop a business proposal showing exactly how XYZ affiliation can save your organization work and money."

The tentative title for the presentation is:

"The Bottom Line"

Development

The theme will be developed in four major steps:

1. The presentation will explain that the key to earning a profit from credit card services lies in the ability to respond to the needs of card holders, merchants and an organization's internal accounting, cost control, fraud control and marketing functions.

2. Next the presentation will describe the two principal ways an organization can handle its credit card operations.

3. From that the presentation will move to an examination of the advantages of choosing a local credit accounting service versus an out-of-state organization.

4. Finally the presentation will highlight the capabilities of XYZ Credit Accounting Services.

The Treatment

The completed project proposal goes to the client. After approving, with or without some modification, the general approach indicated in the proposal—but before giving a final go-ahead for production—the client will probably ask for more specific details on the proposed development of the presentation. Your response to this request is called a "treatment."

If you consider the project proposal to be like an architect's rendering of a proposed structure, then the treatment is like a set of blueprints. It **specifies** how you'll do what the project proposal said you planned to do. It **specifies** the materials you'll use and the directions you'll follow to turn the theme, premise, and content outline of the proposal into the final script and visuals of the presentation.

The treatment, then, interweaves an explanation of the logical development of the presentation's content with a description of the presentation's visuals and sound. The sample treatment on page 105 shows how this interweaving takes place.

The treatment serves two functions. It enables your client to "see"—to mentally picture—your ideas full-blown. As the client reads, he or she should be able to envision the completed presentation. Not that the client will be able to visualize exact images, sequences, and programming effects, but he or she should be able to mentally picture how the ideas and visuals of the presentation will flow together.

The treatment also enables you to see the presentation in your mind's eye. When you complete the treatment, you should be able to "see and hear" the segments of the presentation coming together. You may envision specific sequences—the slides, screens, and visual effects you'll eventually use.

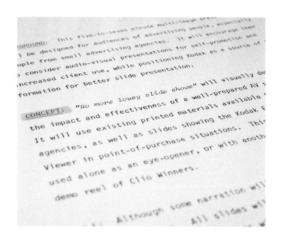

Who Writes A Treatment—And How

The job of writing a treatment should fall to the writer/producer. The slash indicates that this may be one person or two. But whatever the number, the contribution of each specialty should be reflected in the treatment. A multi-image presentation isn't a writer's medium; but neither is it a photographer's, a director's, or a producer's medium. It is the blending of sight **and** sound, of word **and** picture, of writer **and** visualizer. This being the case, the key people who will contribute to the writing and production of the presentation also should contribute to the treatment.

Of course, the scriptwriter may actually pull together the contributions of the production team and express them in the written treatment. But you don't **need** to be a writer to create a treatment. All you need to know is what you want to accomplish and how you plan to accomplish it. Transfer your ideas to paper and you'll have written a treatment.

To write a successful treatment, you must sell your ideas. And to sell your ideas, you have to believe in them first.

So the treatment should reflect your conviction that the approach you've taken will assure success. If you're not totally convinced yourself, then rethink your ideas. Never include in a treatment any ideas or approaches you have doubts about; sooner or later those doubts will resurface as problems with the script or the presentation itself. So, believe in what you're doing and put that belief on paper. If you do, you'll have an effective treatment.

BACKGROUND: This five- to seven-minute multi-image presentation
will be designed for audiences of advertising people, especially
people from small advertising agencies. It will encourage them
to consider audio-visual presentations for self-promotion and
increased client use, while positioning Kodak as a source of information for better slide presentation.

CONCEPT: "No more lousy slide shows" will visually demonstrate
the impact and effectiveness of a well-prepared AV presentation.
It will use existing printed materials available from advertising
agencies, as well as slides showing the Kodak Ektagraphic audio-
viewer in point-of-purchase situations. This module could be
used alone as an eye-opener, or with another component, such as a
demo reel of Clio Winners.

STYLE: Although some narration will be used, music will carry
much of the show. All slides will be masked to the 17 mm by
34 mm ratio and all six projectors will be aimed at the same screen,
providing some animation in a 1:2 format.

STORY: The story will be divided into two parts: the wrong
approach and the right approach.

The treatment can serve an additional function. In a previous chapter, you learned that one method of developing visuals for your presentation was to create a storyboard. If, in fact, you plan to use a storyboard to both communicate and guide your efforts, the time to begin its development is immediately after you complete the treatment.

A treatment **explains** the major sequences of a presentation. It states what will be seen and heard as the theme of the presentation is developed. In some cases, these verbal descriptions are sufficient. But in other cases clients want more specific details; they want to see exactly the angle at which you'll shoot the slides . . . and the composition of the slides' elements . . . and the backgrounds you'll use . . . and the colors you'll employ . . . and on and on. In these cases, you'll have to turn to a storyboard.

While a treatment explains, a storyboard **shows** specific sequences. The amount of detail you'll use to indicate the sequence depends on your needs, the requirements of your client, and the constraints of your budget and time. A sequence can be illustrated with a single, simple sketch, as shown in the illustration on page 107. Or the sequence can be shown in slightly more detail, with the beginning, middle, and ending viewpoints indicated — as illustrated on page 107. This degree of detail is usually sufficient; even the most visually unsophisticated client can fill in the "gaps" between the panels. Sometimes, however, especially when the subject matter is sensitive (such as during the introduction of a major new product), a storyboard may have to be developed even further. Every change of camera angle and position, every change of subject or setting, may have to be illustrated. Only in this way can the client approve or disapprove of your suggested visual treatment.

In addition to the visuals, a storyboard also contains phrases or sentences to indicate the precise point in the treatment — or script — where the visual sequence will appear. The storyboard on page 107 illustrates this verbal and visual keying.

How comprehensive and polished should the panels in a storyboard be? As detailed as understanding demands and time and money permit. If you're working with an in-house client who understands production techniques, the storyboard can be as simple as stick figures you draw yourself. As long as they convey your ideas, stick figures are effective. If your client is with an outside organization, you'll probably want to create a more detailed and professional-looking storyboard. Of course, the professionally rendered board may be no more effective than stick figures in conveying your treatment, but there are times when the medium of a presentation is as important as the message. In these cases, it's better to spend the time and money to create comprehensive storyboards.

Riding The Wave Of Momentum

Although a lot of work lies ahead, when you've completed the treatment you've reached the crest of your developmental efforts. The rest of the effort you put into the presentation, although long and sometimes involved, should feel as if you were riding a wave. The work that remains may require more hours and more labor than you've used in planning, but those hours should flow more smoothly because of what you've done. From here on, you should feel the force of momentum behind you as you work. And that momentum will come because your treatment is a springboard to action; it launches you into the actual production of a presentation.

A storyboard can consist of simple sketches.

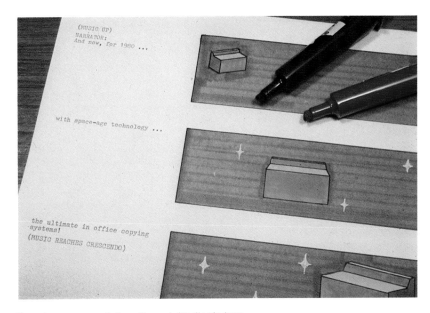

Key phrases or words from the script indicate how a sequence's visual development corresponds to the flow of narration.

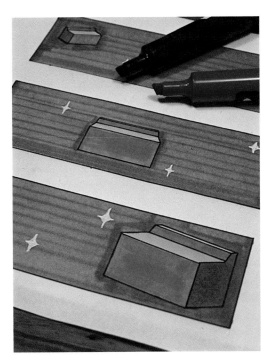

Or it can be more elaborate, showing how a sequence will develop from beginning through middle to end.

Laurence Deutsch

Laurence Deutsch: Planning With Control

All multi-image producers walk a production planning tightrope, says Laurence Deutsch, president of Laurence Deutsch Design, Inc., of Los Angeles.

"On the one hand, you're committed to deadlines. You have to deliver a proposal, a script, or a completed presentation to your client on agreed-upon dates. On the other hand, it's extremely difficult to schedule the many activities that make up a creative project. You can't tell a writer he must write 15 pages of acceptable script every day. Nor can you tell a photographer he must shoot 85 usable slides every day. All you can say is, 'We've got a deadline. Meet it however you can.' "

The solution to this planning dilemma, Deutsch says, is tight control, good communications, and total trust in the people working for you. "You have to focus your resources where they're needed, when they're needed. Then you have to rely on your people. In the final analysis, the commitment to a schedule for creative effort comes from the experience of your people. They know the pressure they're under, and they know how to respond to that pressure."

To focus resources and establish deadlines, Deutsch Design uses a simplified version of the CPM planning method described in this section. They don't create CPM flow charts such as the one illustrated in the Appendix. It would require too much time to create and monitor charts for all the projects moving

Images from a five-screen presentation produced for American Honda Motor Co., Inc. (Photos: Laurence Deutsch Design, Inc.)

through the company during the course of the year. Early in the company's history, however, they did experiment with the method. Its emphasis on the ordering and scheduling of parallel and sequential activities was integrated into the company's planning process, although the actual charting was abandoned, having been absorbed into their thinking.

"Our usual procedure now is to set milestones for major project activities," says Deutsch. "For example, we'll have milestones for the approval of the script, for the approval of the visual concept, for the approval of the site design, for the approval of the presentation itself. But we don't try to set deadlines for intermediate steps. Experience and pressure prods us to meet these subgoals."

Project milestones are established as part of an overall production plan drawn up early in the company's involvement in a presentation. After an initial meeting with a client, representatives from Deutsch Design's major departments meet to discuss the client's needs, goals, and budgetary limitations. Then they relate this information to their own creative goals and profit expectations. The participants in the meeting then take this information and, working alone and together, develop a set of recommendations.

These recommendations, says Deutsch, are based "half on past experience, half on the needs of a specific show. We develop a specific approach, then look at what we've done in the past that might give us indications of the time and costs involved."

As an example, Deutsch points to a "laser tunnel" created for a presentation to introduce the new Mazda RX-7 sportscar. "To begin planning for the time and costs involved in creating this effect, we went back to our experience with lasers in the Mattel Toy Company's **Dimension '78** presentation and several other laser productions. This information helped us plan and schedule production for the Mazda presentation."

The staff recommendations are compiled in a master plan for the project. "Every group is asked to come up with a schedule for the time they'll need to complete a certain project," says Deutsch. "They estimate the time required in hours, then weigh this figure against the demands of other projects already in house. This practice has two benefits. First, we don't get swamped and find ourselves facing hours of overtime. Second, when we tell a client we can deliver on a certain date, we're confident in our commitment."

Next, the master plan is incorporated into a proposal to the client. The proposal contains not only a complete production schedule, but also descriptions covering the scope of the presentation, its length, and the names of the people responsible for various aspects of production. The proposal is presented in a meeting with the client, attended by all the people mentioned in the document.

Out of this meeting comes the information Deutsch Design's creative people use to develop a presentation theme. But more important from a planning standpoint, the client gets to meet the people responsible for the work—and for the deadlines.

"This simplifies and streamlines our operations a great deal," says Deutsch. "If a client has questions about a specific aspect of a show, he or she knows whom in our organization to call. And the reverse of this is that our people can call the client for information when they need it. We've taken out the middleman in these transactions, and that saves time."

Once the creative process begins, it proceeds on several fronts simultaneously. Deutsch Design's facilities people study the show site to determine what they can and cannot do within the presentation area. This information is then given to the group's hardware specialists, so they know what conditions they'll have to work with in setting up and providing power for the projection equipment.

The firm's creative people then begin working with the hardware specialists, to test the feasibility of various visual concepts. While this is going on, Deutsch Design's production coordination group begins making arrangements with the managers of the site, rental equipment agencies, transportation companies, and other outside suppliers.

Deutsch himself coordinates all this activity. "I serve as a consultant to the groups," he says, "but they have total responsibility for their own efforts."

As an in-house consultant, Deutsch draws on a background that reaches into just about every phase of multi-image production. He started his career as an industrial designer, working for a company that created exhibits and presentations for industrial clients. His first major project was an exhibit for the 1964 World's Fair in New York. After that he worked for graphic designer and producer Herb Rosenthal and freelanced for Saul Bass. In 1967 he opened his own company, creating two films and several exhibits for the Queen Mary project at Long Beach, California.

These projects have given Deutsch experience as a writer, photographer, designer, and a film producer and director. But now that his company employs 15 full-time staff members, Deutsch finds himself more of a manager, providing "overall control and leadership" to the group. Still, he finds it difficult to tear himself away completely from the creative

(left) Deutsch Design Producer/Director Kevin Biles programs a presentation working from a show book. (below) Images from **Dimension '78,** a five-screen presentation produced for the Mattel Toy Company. (Photos: Laurence Deutsch Design, Inc.)

side of the business, so he continues to direct the company's film productions.

Not that he doesn't find managing a communications company a creative task in and of itself.

"We take projects from start to finish," he says, "and that involves a lot of work, especially when you consider that we not only produce multi-image presentations, but also live shows and laser presentations. But multi-image is our greatest challenge, because of all the visual media, it's the most complex. It's a very sophisticated medium and you're dealing with more art, more images, more equipment, more screens, more of everything else that you can associate with the visual arts."

Deutsch believes that tight control is the only way to assure success with a medium of this sophistication and complexity, and he points to his company's own record to support his thesis. "We deliver our projects on time and within the established budget," he says, "and that's the key to a production company's success. We've never had to go back to a client for more money, unless, of course, the client wanted to change or expand the original direction of a presentation."

And when that happened, Deutsch could add, he had a plan to deal with it.

109

This section examines the various creative efforts needed to produce a multi-image presentation. Beginning with research and scriptwriting, two skills that form the underpinning of your production, it then goes on to show how the various elements of a presentation are created — art, slides, film, sound. Finally the section explains how all the elements are joined during programming.

Section III
Production

RESEARCH AND WRITING

Getting The Facts

You're ready to begin production. Your plans are set, your theme is established, your treatment is approved. Now all you need is something to say. And something to show.

In short, you need information. You need verbal details—facts, figures, case histories, and quotes. And you need visuals—people, places, events, and objects. And the more the better, because a script—and the final presentation itself—is only as solid as the visual and verbal information used to create it. The best script is a fact-filled script based on intensive research.

Unless you're a historian, journalist, or scientist, research is rarely fun. It consists of sifting through books and brochures, reading newspapers and magazines, locating and interviewing knowledgeable people, and retracing the course of pertinent events. It's time-consuming, mentally exhaustive work. But it has to be done. The opposite approach, "writing off the top of your head," usually results in a script that isn't worth producing.

So if you want a worthwhile presentation, you have to be prepared to do considerable research. Researching can be tedious and demanding while you're doing it, but later you'll find that all this digging for and sifting of information will result in three major benefits:

- **The information you discover will make your presentation more interesting.** Research allows you to talk about people, places, and events. It makes your presentation concrete and specific. Even if your presentation deals with an abstract theme, and even if it presents generalized ideas and conclusions, it can be made more interesting if you support these themes and ideas with concrete details and specific facts. For proof of this statement, just watch TV news magazines such as **60 Minutes.** There you'll find such abstract topics as inflation, the energy shortage, and penal reform treated in a lively, concrete way. The reason: On these programs, abstract topics are discussed in terms of people and events—in terms of specifics.

- **The information will make your presentation more credible.** People are more likely to believe ideas supported by facts and figures. A general statement, left unsupported, fades from the memory of your audience. But take a general statement and support it with facts, statistics, a case history, or quotes from an expert or authority, and that statement will probably be believed and remembered.

- **The information will make your presentation more persuasive.** In a sense, this benefit derives from the first two. If your presentation is interesting and believable, it is probably persuasive too. The opposite type of presentation, one that's abstract in tone and general in content, will likely leave an audience unconvinced of your message.

Three Sources of Information

Where do you find the facts and illustrations that breathe life and power into your presentations? Well, if you're writing a script covering an area in which you're an expert, you don't have to go any further. Your learning, your experiences, your notes, books, and research files should provide you with the facts and examples you need to write a persuasive script. You can skip everything else in this chapter and begin to write.

If you're not an expert, if you're someone who "doesn't know what he doesn't know" about the material to be covered, you can turn to three sources of information for help.

1. Printed materials. This category of information includes books, encyclopedias, newspaper and magazine clippings, annual reports and brochures, business and government studies and surveys, copies of speeches, public relations releases — just about any document dealing with people, products, or events that have made news in the past. Also to be included in this category are the scripts for movies and other audiovisual presentations.

The first source for background information on a subject is usually printed materials.

These materials provide you with background information. They give you names, dates, examples, illustrations, anecdotes, case histories, and quotes. They may also provide explanations and interpretations of events, processes, or theories.

When you're searching for these facts, you're following a lesson from journalism. You're looking for the 5 Ws and the H: who, what, when, where, why, and how. With this information, you have a solid grounding in the subject you'll be writing about. You should also have the beginnings of a framework for your script, plus leads on people you'll want to interview to expand your store of information.

Interviews provide first-hand information.

2. Interviews. Armed with your background information, you're now ready to start interviewing people with first-hand knowledge of your subject matter. These people may be true "experts," those whose background and know-how qualify them to speak with authority about your subject. (Foremost among these experts, of course, may be your client.) Or they may be knowledgeable "participants" or "observers," people who, because they have used or sold a product, visited an area, or viewed an event, can speak in some detail about your subject.

You may be wondering why these experts with first-hand knowledge of a subject haven't been listed as initial sources of information. Isn't it best, you might ask, to start with an expert? The answer is "yes" if you have knowledge of the subject already, "no" if you're still filling your store of background information.

Rushing into an interview at too early a stage in your research usually yields limited dividends. To repeat a phrase used earlier, if you don't know what you don't know, it's hard to ask intelligent and productive questions — those that will elicit the facts and details you need to write a persuasive script. So the best way to approach interviews is to be prepared with questions based on your initial research.

These questions should be phrased to help you investigate areas not fully covered in your previous research, expand areas only partially covered in printed materials, and round out, with new information, areas already touched upon in your initial research. This sort of probing brings your stock of information up to date and completes it, giving you a full understanding of your subject.

Interviewing also performs another useful function: It provides you with quotes that can be used in your script. A person's words are a powerful means of support. They are like the courtroom testimony of an eyewitness. So when you find a knowledgeable, credible, articulate authority or witness, quote him or her whenever and wherever it's appropriate. If possible, you should also arrange to interview these people on audiotape or in front of a camera. If an expert's words are strong, his or her voice saying them is even stronger. Add the person's presence to the voice and you have the most convincing form of support available.

3. Personal investigations. Your background research and interviews should give you most of the facts and examples you need to write a script. Before you begin writing, however, complete your research by examining your subject with your own eyes and intelligence. This means immersing yourself in your subject — visiting sites important to the script, observing people and operations that will be mentioned, using products, and taking tours.

This seeing-for-yourself phase of research is invaluable when writing a script. It may not produce any additional facts — although this is highly unlikely — but it will give you a better feeling for and understanding of the facts you've gathered. This alone will make your writing more interesting and more believable. It will also make writing the script easier. •

Get The Picture

The previous pages have concentrated on gathering the facts, figures, and case histories you'll use to write the verbal portion of your script. Unstated but implied was the importance of finding appropriate visual details to complete the script.

While gathering facts, you also should be gathering visual impressions. You should be making notes about the people, places, events, demonstrations, and objects that will **illustrate** the facts you want to convey. If, for example, you're gathering information on the operation of a manufacturing line, you should also be noting descriptions of visuals you can use to depict this aspect of your story. In many ways, these visual impressions of your subject are more important than the verbal because a multi-image presentation is predominantly a visual medium. You'll want to learn, through reading and interviews — through the use of words — how the manufacturing line operates. But you'll want to convey this information as much as possible through visuals. To communicate this intention in your script, you need a full stock of visual impressions to choose from.

You can record these visual impressions in two ways. One is simply to make note of them, describing as fully as possible the scenes you want to show on the screen. You can supplement these descriptions with notes about possible camera angles and shifts of viewpoint.

Another way to record these impressions is to take photographs of key people and areas to be shown during the presentation. This approach is particularly useful if **someone else** will be returning to shoot the slides to be used in the presentation. These initial "research" shots will give the producer or photographer an idea of the overall setting and photographic possibilities for each subject.

Candid "research" photographs (top) can help the writer, art director or photographer plan for actual shooting sessions (below).

WRITING THE SCRIPT

Audiovisual scriptwriting is a process in which you take a set of words that tell a story and combine them with visual descriptions that tell a story and wind up with a presentation that tells the same story, but more effectively than words or pictures alone.

So what is it that the scriptwriter creates? Not the show itself. Not even the directions for the show. Perhaps an audiovisual script can be described best as "suggestions for the director of a show."

That may seem like a harsh judgment of this specialized sort of imaginative writing, but nevertheless it's true. The fact is, audiences never see a script and they don't want to. They want to see the presentation itself, the final product, not the starting point for production.

So a multi-image script loses its importance as production moves ahead. But at the beginning it is the lifeblood of the entire presentation. The script demonstrates that a multi-image production will work. It shows how the presentation will prove the message stated in the theme. If it doesn't do that, a script is worthless. It may be brilliantly organized and beautifully written, but none of that matters if it doesn't prove your presentation's main point.

The script, then, is the foundation upon which a multi-image presentation is built. It outlines the scope of a presentation, and within that scope it assigns emphasis. It gives life and meaning to the presentation's theme. And it launches all further production. The script, in short, defines and delineates the presentation. It is to the final presentation what a composer's score is to a symphony.

Given its importance, a script must be written with imagination and thoughtfulness, according to specific requirements. It is beyond the realm of this book to teach the methods and techniques of audiovisual scriptwriting. Others have devoted entire texts to that subject. If you want additional instruction in this specialized field, you would be wise to find copies of these books and read them from cover to cover. On the other hand, if your aim is not to become a professional scriptwriter, but merely to write an occasional script, you might find the following suggestions helpful.

1. Think visually. In school you were taught to write for readers. To accomplish this, texts and teachers instructed you to describe scenes, settings, and objects; to narrate events and operations; to explain the meaning of ideas through the logical arrangement of details and examples. You were taught to frame your thoughts in paragraphs that began with a topic sentence and that were developed through the use of methods of support. These are the teachings of rhetoric and they're practical and productive if you write for the print media.

But if you write for audiovisual presentations, these lessons will only lead you astray. That's because audiovisual scripts require you to think and write "visually." In multi-image scripting, you write for viewers and listeners, not readers. And to write for "eye and ear," you have to forget virtually everything you learned about rhetoric. Instead, you have to let "the pictures do the talking" as much as possible.

To let pictures talk, you have to learn a new type of thinking. When you write for readers, you translate your ideas into words. But when you write for visual presentations, this thinking process must be altered. In general, the process of thinking visually proceeds in three steps.

First, you mentally formulate the idea you want to communicate. Let's say you wanted to tell your audience that "inflation is eating away your paychecks."

VISUAL	AUDIO
12. Movie reel cuts on screen. Scenes flicker between the spokes.	Movies bring realism. They help sales people be remembered . . . and create impressions that paper and ink never could.
13. Top view of slide tray empty.	Slides offer flexibility. They're powerful, dramatic, easy to make, and easy to use.
14. Full slide tray, shot in register, wipes around the screen.	
15. Circular marquee, projected from four projectors, borders the screen and flashes. This is filled with the "iris mask" which contains images from New Orleans pictures that look like a multi-image show projected on a circular screen.	And multi-image is a powerful tool for exciting communication . . . because its possibilities are as endless as your imagination.
16. One by one, different screen areas of the multi-image show change to large masked shapes of: a microphone, remote control projector switch, slide, and frames of movie film. Each of these fills with a picture of a person communicating. These, in turn, change to close-ups of MP&AV booklets, programs, etc.	We at Kodak applaud you for your success in professional communications. We share your ideals and your commitment; and we stand ready to assist you.
17. A small circle containing "Kodak does you proud" begins in the upper left quadrant and wipes around the screen becoming larger until it fills the bottom left quadrant. To musical stings, a large "Kodak does you proud" appears in each quadrant. On the final sting, a strobe wipes the screen.	In your pursuit of excellence, Kodak does you proud. [Music up and out]

Second, you mentally select a sequence of visuals that depicts or illustrates your idea. The topic of our example—inflation—is abstract, and an effective multi-image show is concrete and specific. So you must select visual examples of "inflation." You might select scenes showing a housewife shopping, a couple looking for a new house, students entering the gates of a college, or a person in a hospital bed. You might then decide to add graphics to this sequence of slides, showing how costs for essential goods and services have risen over the past ten years. These visuals **show** where and how much inflation is hurting the average consumer.

Third, you write a segment of narration—if needed—to supplement or emphasize the visual idea expressed on the screen. In our example, you might choose not to add words. The visuals and graphics say emphatically that "inflation is eating away paychecks." But if you wanted to supplement the visual message, you could add a quote from an economist or a government official. You might write: "Says government spokesman John Smith, 'Continued inflation threatens the American way of life.'"

That's the process: idea . . . visual interpretation . . . verbal emphasis. For an additional example of this process at work, read the sample multi-image script on page 117. This script shows how the writer told the main part of the story through pictures, using words to supplement the ideas presented on the screen.

2. Write for the ear. When you write to be read, you have certain advantages. You can use paragraphs and punctuation to help convey the meaning of your words and the progress of your thoughts. You can underline or italicize for emphasis. You can break up your material into easily assimilated chunks, using chapters, subheadings, even footnotes.

But you don't have these advantages when writing a multi-image script. You can't hear a paragraph break or a boldfaced word, nor can you interrupt the flow of a presentation for the oral equivalent of a footnote. Nor do listeners have the luxury of time for reflection that readers enjoy. They're forced to hear—and comprehend—speech delivered at about 125 words a minute, with no time to review or to ponder what's been said.

So you have to adjust your writing style—or more correctly, you have to adopt an entirely new writing style—when writing for the ear. You have to write in short, straightforward, easy-to-understand sentences. No long or involved sentences. No unusual or obscure words. No involved style.

The best way to check the appropriateness of your style is to read your script aloud. If possible, record your reading and play it back. This acts as a double check. If you can read it aloud without difficulty, so can a narrator. And if you can understand it as you listen to the audiotape, then your audience should understand it too.

If you discover sentences that are hard to read or to understand, rewrite them. Take the idea you've expressed and recast it in the most straightforward prose possible. Just ask yourself, "What am I trying to say?" When you can answer that question in a single, simple sentence, put it down on paper.

VISUAL	AUDIO

1. To the sound of the musical stings, four triangular-shaped neon designs pop on the screen, one in each corner. Four of the same designs—but in a different color—are alternated with the first designs, creating a spinning effect in each corner.

   [Upbeat music that begins with musical stings]

2. As these graphic designs continue to alternate, one changes to a crudely drawn representation of the same shape. This alternates with the original design. The original design changes to a child's drawing of a stick-figure person, which alts with the crude drawing. Then the crude drawing also changes to a child's drawing of a person and the two-person drawings alt with each other. Up to this point, all screen effects will be created in black to give the illusion they are swimming in limbo.

3. As the narrator speaks, the children's drawings of stick-figure people will change into other examples of children's art.

   Visual communication can be a wonderfully simple process. A child can mix patterns and patches of color into an expression that communicates . . . if not with clarity, at least with imagination.

4. Miscellaneous examples of children's art continues, ending with children's drawings of objects of communications: telephone, pen, movie film, pictures, etc.

   [Music up]

5. In a series of dissolves, the children's drawings change into more refined drawings of the same objects. These, in turn, dissolve into pictures of the real objects, photographed against black.

   Imagination is the gift we all possess . . . and from the beginning, it's a vital part of our communication.

3. Write in sequences. In school, you were taught to frame your thoughts in paragraphs. Begin with a topic sentence, then expand, explain, and support your idea using details and examples.

The principle of paragraphing—statement and support—is also valid in audiovisual scriptwriting. But the form—the structure—of a paragraph should be discarded. By its very nature, it encourages wordiness, a quality that weakens audiovisual scripts.

Instead of writing in paragraphs, you should write in sequences. A sequence is a segment of a script that introduces and examines a new topic—both verbally and visually. Unlike a paragraph, which reveals the progress of a writer's thought as it moves from a general concept to the specific details and examples that support the concept, a sequence presents the general and specific together. At times the specifics are words—the narration of facts or explanations or causes—supplemented by visuals used to create a general impression. Far more often, the specifics are the visuals—people, places, and events of primary importance—that are supplemented by words, the narration of ideas, qualifications, and conclusions. But no matter which element dominates a particular sequence, the general and the specific are united in thought and simultaneous in presentation.

This idiosyncratic quality of scriptwriting requires a writer to think in terms of sequences—units of words and pictures that flow as single streams of thought. Mastering this type of thinking—and writing—takes practice, and it is the lack of this practice that flaws the beginning scriptwriter's efforts. The novice scriptwriter writes what amounts to a prose essay, then tries to affix visual descriptions to the paragraphs. But these visuals matched to the words don't make a script. The product of such a process is usually wordy, rhetorical in nature, and visually disjointed.

So to write an effective script you have to think in sequences, and to learn to think in sequences you have to:

1. watch and listen to as many movies, multi-image presentations, and audiovisual shows as you can, paying particular attention to how words and visuals work together;

2. read as many movie, multi-image, and audiovisual scripts as you can obtain, paying particular attention to how the writer linked the visuals and words in sequences; and

3. write, write, write multi-image scripts until your product approaches the models you've studied.

As an example of what to look for in your study, examine the sample script on page 119. It shows how one writer handled a sequence of ideas. As you can see, words and pictures, statements and support, the general and the specific, are presented simultaneously. The writer is telling the story through visuals. These form the specifics of the sequence. The more general words supplement the visuals, offering an introduction and an interpretation. Notice also that the words, in this secondary role, derive their order and position from their relationship to the visuals.

SEQUENCE OF IMAGES SHOWING SALES REPS WORKING WITH A SERIES
OF TERRITORY PLANNING AIDS: MR&A SCROLLS, ADVERTISING LEADS,
A COPY OF THE WALL STREET JOURNAL, A COPY OF DUN & BRADSTREET,
LOCAL BUSINESS NEWSPAPERS, CONVENTION INQUIRIES, ETC. WE SEE
MEDIUM SHOTS OF THE SALES REPS AND CLOSEUPS OF THE
MATERIAL THEY'RE USING.

> NARRATOR: ... a thorough knowledge of the opportunities
> that exist in their territories. This knowledge
> comes from a number of sources ... market
> research and analysis summaries, newspaper
> reports, advertising leads, convention inquiries,
> and personal research.

SEQUENCE OF IMAGES SHOWING SALES REPS WORKING WITH
CALENDARS, PLANNING GUIDES, POCKET CALENDARS. THESE ARE
ALL MEDIUM SHOTS.

> NARRATOR: Organizing this research into an effective sales
> plan requires day-by-day scheduling — something
> each rep does in his or her own way. Because
> at Kodak, sales reps are encouraged to manage
> their own territories.

SEQUENCE OF IMAGES SHOWING CLOSEUPS OF THE PLANNING
GUIDES, CALENDARS AND OTHER DOCUMENTS SALES REPS USE TO
ORGANIZE THEIR TIME.

The Two-Column Format

A multi-image script—for that matter, any audiovisual script—can be written in one of two formats. The first, called the "Hollywood" format, follows the conventional, horizontal orientation of most written material. Visual descriptions are written from margin to margin and typed in capital letters. The narration and notes describing music and sound effects are written in narrower columns that are centered on the page. (See example on page 121.) This sort of presentation makes it easier for producers, directors, and studio executives to read and approve feature-length scripts.

Most multi-image scriptwriters avoid this format, however. Its major drawback is that the horizontal separation of audio and visual descriptions makes it difficult for a producer to key his or her slides to the narration.

To overcome this problem, multi-image scriptwriters use the two-column format. (See sample script on page 123.) The visual descriptions are typed in the left column, the narration and sound descriptions in the right, adjacent to the corresponding visual sequences. By aligning visual and verbal in this way, the scriptwriter allows the producer to plan the visual presentation quickly and with little confusion.

Revisions

For a scriptwriter, revisions are like taxes and death. They're inevitable, physically depleting, and destructive of the life of the script. If you don't believe so, just ask any writer.

But ask producers about revisions and you'll probably hear them described in opposite terms. Revisions, according to producers, make good scripts even better.

The producers are probably right. The first draft of any script, no matter how well thought out and written, is usually longer and wordier than it needs to be. This being the case, a good producer will want words taken out, sequences shortened, and the pace quickened.

So unless you're a writer/producer who knows exactly what you want to begin with, be prepared to make script revisions. And be prepared to make them in the right spirit—respecting the intelligence and experience of your collaborators.

Pride of authorship only makes the work of revision harder. As a scriptwriter, you must realize that a multi-image production is the product of many skills and talents. So you can't regard a script—your script—as the "last word." It is, at best, a working document, one that will be edited and altered until the narration and sound track are completed.

But that shouldn't be a cause for despair. **The scriptwriter's words are the starting point for production.** It is revisions and collaboration that result in the eventual presentation.

VISUAL	AUDIO
8. The word "excellence" flops to all sides of the screen, then dissolves to a top shot of heavy metal gears and printed circuit boards. Superimposed over these are Wess slide mounts to create the effect of an open-gate slide held in front of a scene. This dissolves to a black-limbo scene of the ends of three projector lenses in the same position as the Wess slide mounts.	The pursuit of excellence is the spirit of professional communication. The professional's goal is to express clearly, concisely, and in a language that all understand. Visuals are such a language.
9. The projector lens shot cuts off and is replaced by visuals that illustrate the concepts: instructive, diverting, analytic, catalytic, and unifying. These are in different formats and have random orientation. As the music comes up, these change and begin a sequence of four business and industry slides. This changes to a medical sequence which, in turn, changes to a classroom situation.	Visuals. They're instructive. Explanatory. Diverting. Analytic. Catalytic. They bring people together, get agreement, and provide a basis of understanding and experience on which organizations and ideas can grow.

[Music up] |
| 10. Visuals of billboard, magazine ad, newspaper picture, graphic storefront, and television set blink on and off, overlapping each other creating a visual mess. Strobe goes off, flooding the screen with light, then leaving it in darkness. | In a world saturated with visual statements, it is not enough to use visuals. [Music becomes a wild cacophony then ends sharply] |
| 11. Four-part slide showing movie reel, slides, audiotape, artwork, script and programming sheet fades up from black. | The challenge is to use visuals well . . . and to remember that each medium offers unique possibilities for effective communication. |

SHOOTING SLIDES AND RECORDING SOUND

The Requirements of Multi-Image Photography

You don't judge a book by its cover, and you shouldn't judge a multi-image presentation solely on the basis of its photography.

But the average audience does just that. They remember what they see — the slides and film sequences you use to illustrate your message. It doesn't matter that these sequences may be the result of a scriptwriter's inspiration or a programmer's imagination. To an audience, script and programming remain out of sight and out of mind. They see only the slides and motion pictures. That's what they react to, that's what they judge, and that's what they **remember.**

This being the case, your first consideration when undertaking photography is the creativity and technical competence of your photographer. Quite simply, you or your photographer should have professional experience in slide photography. If your photographer doesn't, find one who does. If you don't have experience but must, for whatever reason, shoot slides for a multi-image presentation yourself, then put this book aside for a while and study one or more of these basic texts.

This chapter extends a basic knowledge of photography into multi-image production. It examines several areas where you must give special consideration to the requirements of a multi-image presentation.

Professional equipment and talent are essential to produce high-quality media.

124

1. Planning the visual sequences. A multi-image presentation isn't a simple audiovisual presentation with additional images—at least it shouldn't be. When you use multiple images, you should do so with a purpose. Each image you project on the screen should extend the meaning of a sequence—provide another viewpoint, offer another comparison, or concentrate the viewer's attention on an additional detail. This requirement demands more planning on your part.

You can appreciate this need for additional planning more vividly if you contrast the visual development of a single-image presentation with that of a multi-image presentation. In planning a sequence for a movie or a single-image slide presentation, you approach a subject as if duplicating the path traveled by your eyes. You begin with an overall view, then tighten to a medium view, then move in for a close-up, and maybe even an extreme close-up study.

For example, let's say you're shooting a sequence focusing on the efforts of an art director in an advertising agency. You could begin this sequence with a long shot of the building in which the agency is located (while your narration introduces the topic of art direction in advertising and, in particular, the work of one art director in a specific agency). Your next sequence might show a full shot of the art director at the drawing board (while the narration introduces him and cites relevant work experience). Then you might move in for a head-to-waist shot of the art director, with the camera angled to also show the layout he's working on (while the narration introduces the subject of layouts in advertising). Finally, you might tighten your focus to an extreme close-up of the art director's hand as he completes an element of the layout (while the narration relates layout to final ad).

Plan visual sequences to follow a logical continuity.

This approach recreates the visual path your eyes might take if you were walking into the agency to watch the art director work. Of course, in planning the photography for a single-screen audiovisual presentation, you would select only the highlights of this journey. You wouldn't show every step, only those that reveal the essence of your message.

Your audience will follow your steps—and fill in the gaps in your visual journey—if you maintain continuity throughout the sequence. Continuity is the word used to describe the flow of events in a slide or film sequence. If each image in a slide sequence looks as if it flows out of the action and setting of previous images, then your sequence has continuity.

When developing a sequence for a multi-image presentation, you use these same techniques—planning the action and maintaining continuity. But their application must be extended. As mentioned earlier in the book, multi-image presentations allow viewers to see more than the unaided eye can see. So your planning must **extend** your sequences to include more visual information.

You can extend a multi-image presentation's "vision" in three ways.

• **You can multiply the elements included in the action of a sequence.** This reduces the number of gaps the viewer must fill in and strengthens continuity. Applying this approach to the art

Develop each series of images to duplicate the details a visitor's eyes might see if he or she were entering the scene.

director example, you could show not only the exterior of the building, but also—simultaneously—the agency's listing in the building directory, an elevator control panel indicating a car stopped on the designated floor, the entrance to the agency's offices, and a receptionist looking up from her desk. These shots present additional details of the path followed to reach the art director—but without adding more time to the presentation. In effect, you compress time and extend vision.

• **You can multiply the viewpoint of a single subject.** This helps you establish the identity and dimensions of your subject. To go back to the example of the art director, you might, when you reach the point where you want to show him in his office, shoot slides of the director from the front, both sides, from the back, and from below—as if the camera were looking up from the drawing board. This additional information rounds out your portrait of the art director and his work—without requiring additional time.

• **You can multiply the details in a sequence.** This approach helps you characterize the subject and setting of a sequence. If, for example, you shot details of the art director's office—close-ups of previous ads, the books and magazines he keeps in the office, awards that hang on his walls, his work materials, a list of coming assignments—you would be enhancing the audience's appreciation of the man and his work.

Your goal then, when planning multi-image sequences, is to consider both the length and breadth of the sequence. The length represents the basic continuity—the action that will take a sequence from a logical beginning to a logical end. The breadth represents additional information—the multiple images that give a sequence greater detail and meaning.

Keeping Track Of Your Photography

During the course of a multi-image production, you may shoot thousands of slides and tens of thousands of feet of film. This volume of material could easily swamp you if you didn't have a system of keeping track of it.

The easiest way to know "what's what" in your boxes of slides and rolls of movie film is to keep a film log for everything you shoot. The log at the right was used during a film production. Its entries helped the film editor sort out film footage by:

1) scene number and action;
2) good or bad ("NG") takes;
3) silent or sound filming.

With this information an editor can quickly find, from the thousands of feet of film he or she must work with, the precise scene needed to construct a sequence.

(Other entries in the log enable the camera operator to give specific instructions to the processing lab on the conditions under which the film was shot.)

Some producers also find it helpful to keep logs of the slides they shoot. They do this by keying their rolls of film, usually by shooting a slate on the first frame. (This can be as simple as a sheet of paper with a number on it.) Information on the subject and settings for each roll is kept in the log. When the slides are returned from processing, they're kept in the lab's packing carton, which the producer labels from the information on the first slide. Using the log and the labeled boxes, the producer can quickly find slides needed to create a visual sequence.

A system for identifying and logging film and slides makes later sorting and editing easier.

2. Directing. All the considerations mentioned above about planning your visual sequences apply equally to directing the photography. The latter activity is just a step removed from the former. So in directing you must ensure that the intent of the visual planning is carried out. In addition, you have to interpret each scene based on the requirements of your visual format.

This isn't difficult if you're handling photography yourself. Having developed the visual format, you'll know the size, shape, and position of the visual "slots" you have to fill. But if you're directing other photographers — and especially if some of the photographers will be working without

A professional director communicates specific requirements to a photographer.

continuous direction — you're going to have to communicate, in detail, the requirements created by the visual format. This means giving your photographers three sets of directions:

● First, you have to explain the visual format of the show and indicate how the requirements of this format translate into slide aspect ratios. Specifically, you must tell a photographer if you need 24 x 36 mm transparencies or one of the super-slide sizes; if you're using a horizontal, vertical, or mixed format for your slides; or if you're planning to mask certain slides to create special effects. Only with these specific instructions can a photographer shoot the slides your visual format requires.

● Second, you have to explain how the slides will be used. Will they be used as separate images, with each slide independent of those that precede and follow it? If so, the photographer can use a hand-held camera for shooting. Or will the slides be used in an animated sequence or a progressive disclosure? If so, the photographer must use a tripod with the camera to maintain a uniform orientation throughout the shooting session.

● Third, you should convey to the photographer the style of photography you want. This is especially important in multi-image presentations, where a number of images appear on the screen at the same time. If this grouping of images reveals a variety of photographic styles, your presentation will look as if it were produced from slides found in a desk drawer.

The easiest way to convey these three sets of instructions is with a storyboard. Photography is a visual art, so words alone may not fully convey the scope of your instructions. A storyboard, on the other hand, **illustrates** your visual approach. It leaves little to chance and nothing to misunderstanding because the photographer can see how each element fits into the overall presentation.

128

EASTMAN KODAK COMPANY	SUBJECT_____
MODELING • PERFORMING	PRINT
NARRATION RELEASE	PROJ. # _____
	HRS. WORKED

For value received and without further consideration, I hereby consent that all photographs taken of me and/or recordings made of my voice or musical performance

at _____ on _____ 19_____

by _____ for the Eastman Kodak Company, may be used by the Eastman Kodak Company, or/and others with its consent, for the purposes of illustration, advertising or publication in any manner.

SUBJECT _____
SIGNATURE

SUBJECT'S SOC. SEC. No.

Street_____ City_____ State_____ ZIP _____

IF SUBJECT IS A MINOR UNDER LAWS OF STATE WHERE MODELING IS PERFORMED

GUARDIAN _____ GUARDIAN_____
SIGNATURE   PRINT

Street_____ City_____ State_____ ZIP _____

Date _____

The Model Release

With the exception of individuals who have chosen a profession or lifestyle that keeps them in the public eye, people have a right to privacy that you must respect when producing a multi-image presentation. To avoid potential lawsuits arising from the unauthorized use of someone's picture or voice, you should obtain written permission from anyone you photograph (on slides or movie film) or record if the individual's face or voice will be recognized in a presentation to be shown to the public—regardless of whether or not admission is to be charged.

You don't have to obtain this permission at the time the slides are shot or the audiotape is recorded, but it's usually easier if you do. If you wait until after selecting the slides or editing the audiotape that will be used in the presentation, you must retrace your steps to obtain permission. This can become a time-consuming, inconvenient, or even impossible task, especially if you shot slides or recorded voices at a number of locations.

The simplest way to get permission is to have your photographer, sound recordist, or their assistants ask each person photographed or recorded to sign a model release (similar to the one shown above) prior to or at the time of shooting or recording.

Finding The Right Photographer

A photographer is a photographer is a photographer. Unless, of course, you're talking about shooting for a multi-image presentation. Then the matter of previous experience should be considered.

What sort of experience should you be looking for? Well, if you've never hired or directed a multi-image photographer before, you might overlook some basic criteria.

1. Your photographer should be accustomed to taking multiple shots of the same subject and to shooting his or her slides from a variety of viewpoints. One or two shots of a subject won't do you much good if your presentation requires a dozen or more slides of that subject—with enough variation among the slides to avoid monotony.

2. Your photographer should know how to tell a story with pictures; that is, he or she should see sequences in terms of a beginning, a middle, and an end. This way of looking at a photographic session is somewhat different from that of a newspaper or advertising photographer, who looks for one dramatic picture—the single image that tells the entire story.

3. Your photographer should be competent using a 35 mm camera. The reason is obvious: Most multi-image presentations use 35 mm slides, but not all photographers work with 35 mm cameras. So make sure the photographer you choose can produce slides in the format you need.

4. Your photographer should be able to work quickly. Some highly competent professional photographers work in a deliberate and controlled fashion. Unfortunately, these qualities, while acceptable in other fields of photography, can spell disaster in multi-

image production where prolonged shooting sessions can mean excessive budget overruns. So choose a photographer who can work competently **and** fast.

5. Your photographer should be able to work on location, using available light or portable lighting. And to reemphasize what was said in the previous paragraph, when your photographer must use portable lighting, he or she must be able to set up and adjust this lighting quickly and professionally.

6. Your photographer should be able to direct people during a shooting session. This requires more than the ability to pose people; it requires the ability to work **with** people, to put them at ease, and allow them to act naturally.

Given these criteria—but allowing for exceptions—you can usually judge a photographer's suitability for multi-image photography on the basis of experience. At one level of experience is the photographer who can shoot professional-quality slides, but who lacks experience in audiovisual production. At another level is a photographer with experience in shooting for audiovisual productions, who understands the problems of shooting for visual continuity. Finally, there are photographers who have shot slides for multi-image presentations. They know the problems of multi-image production and shooting for a specific visual format; they also know how these problems translate into shooting requirements. If you can find a photographer with this kind of experience, you'll simplify the job of directing. If you can't find such a photographer, you should expect to spend more time directing the shooting sessions.

3. Shooting Panoramas. A panorama is a wide view of a subject, filling the full screen area and generally created by projecting two or more slides simultaneously. The projected slide images are often separated by narrow vertical spaces. A panorama can also be created with one slide by using a projector with a shorter focal-length lens, but this may introduce problems in maintaining satisfactory image brightness.

Like all dramatic effects, panoramas should be saved for a special moment in a presentation, one in which you want to gain maximum visual impact. So like a magician who keeps the best trick for last, you should keep the panorama in your catalogue of effects until that precise time when communication requires its use.

You can shoot slides for a panorama in a number of ways:

• The simplest method is to set up a 35 mm camera on a tripod, being sure it's level; then aim the camera so that the right edge of the frame is at the precise midpoint of the scene you want to shoot.

Then, after taking several exposures, swing the camera to the right so that the left edge of the frame lines up with the same image midpoint. Shoot several exposures at this point, being certain to cover the same exposure range you used for the first shot. The resulting slides will make a two-screen panorama. If you want to make a three-screen panorama — or more — simply duplicate the action described for each additional screen.

Another point to keep in mind when shooting this type of panorama is to have the camera pivot under the optical center of the lens, rather than back under the camera body. This will provide a better image edge match. The optical center of the lens can be found accurately enough by measuring a distance equal to the lens focal length **in front of the film plane.**

• A more precise method of creating this type of panorama relies on an indexing head mounted on a tripod. Instead of using points of reference in a scene to align your camera, you use the calibrated markings on the indexing head. Most such indexing heads are also calibrated for use with lenses of different focal lengths. Here again, it is best to pivot the camera on the optical center of the specific lens being used.

• You can also use a panoramic camera to create your slides. These special cameras advance film past the lens as the camera is swung through the panoramic field of vision. The film is returned from processing in a strip, which must be cut and mounted by hand.

Panorama created by butting two images.

Indexing-tripod heads simplify shooting panoramic images.

If You Use Motion Picture Film . . .

Motion picture film adds still another dimension to a multi-image presentation. Its use also adds another dimension to your planning.

Your first planning consideration, of course, should be whether or not you need motion pictures. Integrating film into a multi-image presentation compounds your work, not only during photography and editing, but also during programming and presentation. So don't use motion picture film just to use it. Have a specific reason, such as a sequence that would not be as effective without actual motion.

This might be the case if you wanted to illustrate the operation of a specific piece of equipment or the motions of an athlete or dancer, subjects that require real-time motion. You might also use film clips for interviews, where it's important to show people to heighten the credibility of a sequence. Or you might use motion picture film to gain the special benefits it alone can offer, such as the use of high-speed photography (slow motion) to analyze operations.

A second planning consideration when shooting motion picture film is the placement of the images on the screen. Your visual format determines where on the screen your images will fall. So your cinematography must be planned and shot to fit into these screen areas. That means you'll have to give special consideration to camera angles and composition of the images.

If, for example, your motion picture images are going to appear on the left side of your screen, you don't want to shoot a sequence in which the action is moving to the left, away from the other images. Nor, for the same reason, do you want to photograph people who are looking to the left. The only way you can avoid these problems is by proper planning.

• You can achieve a panoramic effect — with less perspective distortion than the three previous methods — by shooting your scene with a 4 x 5-inch camera. Shoot two transparencies of the scene — being careful to keep the camera locked in the same position and using the identical exposure for both transparencies. When the transparencies are returned from processing, you cut 35 mm slides out of each of the two larger pieces of film, the left half of your two-screen panorama from one and the right half from the other. Be sure to allow about 1/16-inch (1.6 mm) overlap between the two slides to provide for the masking edge in the slide mounts. If you're planning a three-screen panorama, the left and right screen images should be cut from one of the 4 x 5 transparencies and the center image from the other. Again, by cutting the slides in this pattern, you avoid leaving a cutting edge gap in your panorama when the slides are mounted. (This method also can be used when the original slide is shot on 35 mm or a 2¼-square format and then duped to 4 x 5 inches. The disadvantage of this approach, however, is the increased grain that occurs in the enlargement.)

• You can use a tripod-mounted platform to hold two or three identical 35 mm cameras having the same focal-length lenses and same shutter speeds, positioned so they provide overlapping fields of vision. Using shutter-release cables, you trigger the cameras simultaneously. This technique is especially useful when you're shooting a panorama containing people or objects in motion.

• You can use one of the relatively new, superwide-angle cameras. Two identical transparencies from one of these cameras would allow you to produce either a two- or three-screen panorama.

Creating a panorama with seamless masks: The image as it appears on a full-frame transparency.

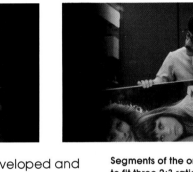

Segments of the original transparency are duped to fit three 2:3 ratio slide mounts, then sandwiched with special seamless masks.

On the screen, the three images blend to form a visually dramatic panorama.

• You can use a recently developed and simple approach to creating two- or three-screen panoramas. Shooting the original transparency for this method is the same as those listed above for the 4 x 5 camera, the panoramic camera, the 2¼-square camera, or the superwide-angle camera. However, instead of mounting the segments of the original transparency or enlarged duplicate to "butt" with each other on the screen, the segments — or duplicates of them — would be sandwiched with continuous-tone masks. These masks — called seamless masks, soft-edge masks, or shadow masks — create a gradual blending of one portion of the image into the next one with virtually no visible seam between the two. However, to achieve a two-screen panorama with this technique actually requires three images (and therefore three projectors) for each panorama, unless you allow for a sufficient overlap of your two screen areas. Of course, this technique also can be used to create a three-screen panorama. Details on the creation of these masks are located in the section entitled The Magic of **Kodalith** Film, page 172.

Bob Kirchgessner

Bob Kirchgessner: Using Analysis To Create A Visual Style

When Bob Kirchgessner talks about developing a visual style, he doesn't discuss creative imagery.

At least not at first.

"Analysis of your goals, audience, communication objectives, site and budget must come first," says Kirchgessner, general manager of Lighthouse Productions of Cincinnati, a service of Scripps-Howard Broadcasting Company. "Then, and only then, can you begin to think creatively about the development of your visual style and your visual sequences."

Kirchgessner's emphasis on analysis before creativity is an outgrowth of 18 years of experience with multi-image production, including work on presentations for the Seattle World's Fair, the Paris Air Show and countless industrial clients. He started his career as a specialist in electronics, an inventive artisan who modified early audiovisual equipment to meet the more demanding requirements of multi-image presentations. He later traveled with road presentations, taking them to city after city until, as he puts it, he "learned to cope with chandeliers, inadequate power supplies and many other site problems." Realizing he also had a knack for creating the materials to be used with his equipment, he then moved into production.

In planning meetings, Kirchgessner, creative director, and producers brainstorm the development of a visual style for current productions. (Photos: Lighthouse Productions)

His varied background taught him "how to make multi-image work." The most valuable lesson learned was the importance of doing a complete analysis of a client's needs and goals prior to the development of a creative course.

"A successful multi-image presentation communicates a very specific message within a very specific framework," he says. "The key to creating a message lies in precisely defining the framework."

At Lighthouse Productions, the time to define the framework for a new presentation is during a brainstorming session, called by Kirchgessner before any work begins on the production. In addition to Kirchgessner, the sessions are attended by Lighthouse's creative director, four producers and the director of marketing. During the session, they analyze the client's requirements to define specific needs: What is the objective? Who is the audience? Where will it be presented? What will the budget allow? Once they've analyzed their objectives, subject, audience and resources, the participants brainstorm their way to an overall program concept.

That concept is then assigned to a scriptwriter who, working with the assigned producer, writes the script for the show. The script contains both the presentation's narration and a description of visual sequences.

A client's experience in audiovisual production normally dictates the next step in Lighthouse's visual planning. If the client has experience with multi-image techniques, the descriptions in the script are usually sufficient. However, if a client lacks multi-image experience, Lighthouse's creative director prepares generalized storyboards to illustrate highlights in the development of visual sequences.

Once the client approves the visual treatment production moves into the company's art and photography departments.

The next step in Lighthouse's visual planning process is to test their ideas on the screen.

"We take a two- to four-minute sequence, develop the visuals, then program the sequence to the music we intend to use in the show," says Kirchgessner.

If necessary, the visuals, screen movement and timing are modified, then the sequence is shown again. The producer continues testing the visual concept in this way until he or she is satisfied with its total effect. On the basis of changes made during these preview screenings, the producer will modify the entire visual plan for the presentation.

With some productions—those for major presentations or those for clients having a difficult time visualizing the final product—Lighthouse also invites the client to see the polished version of the test sequence.

Lighthouse used this preview approach during the visual planning and development for a presentation produced for Expo '82.

"This was a major production," Kirchgessner says, "involving multiple screen effects. No storyboard or script description could fully illustrate or depict the effects we were attempting to create. So we demonstrated to the client—and to ourselves—just how the visuals and programming would be used to produce a key sequence."

For other clients, especially those with little or no experience in multi-image production, Lighthouse often uses its own presentation, entitled "Creativity in Communication," to judge receptiveness to multi-image techniques and effects. "It's been a very productive show," Kirchgessner says of the presentation. "First of all, it sells the idea of multi-image to a client, by creating excitement in his mind concerning the medium. Second, it helps us gauge the level of sophistication the client seems most comfortable with."

The presentation, of course, is a further refinement of Kirchgessner's use of analysis to fuel the creative process. It provides him with an objective appraisal of a potential client's communications awareness. That appraisal gives him one more bit of information with which to develop visual concepts.

Needless to say, it also solidifies his belief in his creative methods.

"Our approach is based on 18 years of experience," he says. "What we try to do is adapt what we have learned in the past, what's worked in the past, to the specific needs of a current client. Our planning system places the goals of our client first. And, by doing that, becomes more productive, dramatic and effective in terms of content and creativity."

Then, emphasizing his pragmatic approach, Kirchgessner adds: "And, we know it works."

Facilities and presentation crews prepare presentation site.

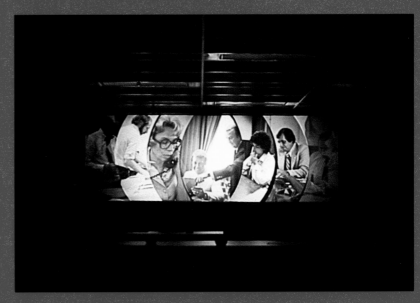

A presentation produced for United American Bank illustrates the use of masking to turn an otherwise commonplace series of "people shots" into an interesting visual effect. (Photos: Lighthouse Productions)

135

Who Needs An Art Director?

A multi-image presentation is, for the most part, a photographic medium. So who needs an art director? Probably the person who would ask that question. People who produce multi-image presentations come from a variety of disciplines: photography, cinematography, writing, directing, advertising, marketing, education, training, music, and on and on. The point is, in most cases a multi-image producer has strength in one or more of the areas. And for the areas in which he or she doesn't have expertise, a specialist is usually called in.

One of the most frequent exceptions to this practice includes graphic design. Unfortunately, not all producers realize that art direction is a specialty that demands a specialist. Nor do they realize that the value of an art director to a multi-image producer goes far beyond the execution of an illustration or a cartoon or a bar graph.

Consider for a moment the visual appearance of a well-designed catalogue or brochure. When you look at it, the photographs, diagrams, art, and headlines all fit together. They complement each other.

Now consider the total screen area for a multi-image presentation. When you start projecting images onto that screen will they "fit together"? Or will they just be a random display of visuals? And how about the type style you use? Will it be legible? Will it be dynamic and exciting? Or will it just sit there?

Just as the pages in those brochures don't simply happen, the aesthetics of what you put on a screen or screens won't be right automatically. It takes a sense of design. And if you don't have this sense, you should make certain someone on the production team does.

Now that we've recommended that you approach your presentation as an art director approaches the design of a catalogue or brochure, we'll retreat one step and throw in a caution. Designers and art directors who work with print media are usually oriented towards static designs— those that can be studied. Multi-image presentations, on the other hand, are far from static, and can seldom be viewed in a leisurely fashion. So unless your designer appreciates the possibilities inherent in a multi-image visual format, you'll have to brief him or her thoroughly before any design work is initiated.

If you're planning to use art for any of the visuals in your multi-image presentation—whether it's to be illustrative, cartoon, diagrammatic, or typographic—follow a basic rule: **Keep it simple.** This is true of art for any audiovisual format, but particularly important for multi-image. Your audience must see and interpret many images in a brief time span, so to be effective, artwork must be basic and uncluttered. If the art is a diagram or a graph, incorporate only the essential points. If it is typographic, use only key words.

The use of typography—for audiovisual shows in general and for multi-image shows in particular—presents an additional problem: legibility. The last thing you'll want is to have copy on the screen that can't be read by those members of your audience who are farthest from the screen. Legibility can be affected by a number of factors, including type style, type size, distance between viewer and screen, and type "weight."

Let's look first at the style of lettering. With the introduction of phototypesetting several years ago, the number and variety of type styles available have grown to the thousands. Some are very exciting visually—but not necessarily very legible when the viewer needs to read them quickly. This doesn't mean that all the type projected on a screen has to be Helvetica Medium or a similarly austere style. It simply means that you should be selective when choosing a typeface for audiovisual use. Naturally, the bigger you can make the type on the screen, the easier it will be to read.

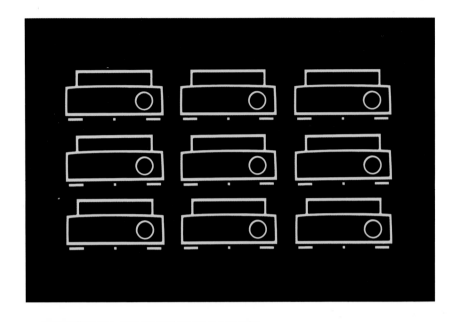

Individual
Employee
Data Needs

137

That leads us to a second factor affecting legibility: type size. This subject has been studied extensively, and a number of complex formulas have resulted—all of which point to one basic guideline: **If you're putting type on a screen, it has to be large enough to be read by the people in the last row.**

Unfortunately, accomplishing this isn't always as easy as saying it. For example, over the years, a couple of rules of thumb have evolved for setting type the proper size. One stipulates that the type size be at least 1/25 the height of the artwork image area. The other simply says that if you can read the type on the slide without a magnifying glass, it's large enough. Although both of these rules of thumb might produce acceptable results, they are just what their name implies—rough estimates.

8 POINT
Printing has performed a role of achievement un
Printing has performed a role of achievement un

16 POINT
Printing has performed a
Printing has performed a

24 POINT
Printing has perfc
Printing has per

A third factor affecting legibility is the distance between the viewer and the copy he or she is trying to read. As mentioned earlier, your audience should be located in an area that is no closer to the screen than twice the height of the screen image and no farther than eight times the height of the image from the screen. It stands to reason, therefore, that if your farthest viewer will never be more than four times the height of the image away from the screen, the minimum size of type need not be as large as when that farthest viewer is eight times the height of the image away.

A fourth factor that influences legibility is the "weight" of the type style and the difference in size between the upper and lowercase characters of that particular style. A very delicate—or "light"—type style will be more difficult to read at any given distance than a bolder style of the same size. Similarly, copy that is set in all uppercase (capitals) is more visible at a given distance than the same copy set in lowercase. Note, however, that we said only it is more visible—not necessarily more legible. Extensive studies have shown that lowercase type is more legible—given optimum viewing conditions. So, the ideal situation would be to have your type set in lowercase, but in a size large enough to make it as visible as uppercase type. This means the height of a lowercase character (without any ascenders or descenders) should be as large as the acceptable size for an uppercase character at the given distance.

The accompanying illustration shows a Legibility Calculator developed by Eastman Kodak Company. It can be used not only for determining the appropriate size of type on new artwork but also for checking the acceptability of type sizes on existing material. (A full-size reproduction of this calculator appears in Kodak Publication No. S-30, **Planning and Producing Slide Programs.**)

Although Helvetica Medium is an easily readable type face, it is not the only type face a producer can choose.

A wide variety of highly legible type faces are available, such as this Gill Kayo transfer type made by Letraset USA Inc.

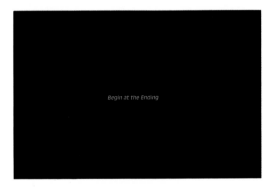

Begin at the Ending

Minimum type size for legibility — but producer / art director could make better use of screen area . . .

RIGHT MOOD

. . . as is shown in this example.

A Legibility Calculator can take some of the guesswork out of determining an appropriate type size.

By using standard-size art on registration boards, an illustrator can save a camera operator a considerable amount of time as well as simplify the storage of finished artwork.

Preparing Art For Copy. Although most graphic arts studios are familiar with the techniques required to produce artwork for slides, the subject is still worth discussing here, at least briefly.

A good place to begin is with the standardization of size. Standardization provides two benefits. First, if all of your art is prepared on the same size board (10 x 12-inch [254 x 305 mm] is a good choice), it will be much easier to handle and file. Second, the shooting of slides from standardized artwork will be simplified immensely because the copy camera won't have to be refocused after each piece of art has been shot. This can be especially important if changes are made on some of the art and those pieces have to be reshot at a later date to match the field size of the initial slides.

If the slide art to be used for your multi-image presentation is relatively simple, or if a member of your staff is artistically inclined, you may choose to prepare some of the slide art internally. A wide range of readily available materials makes the task easier and the product more professional. Most art supply houses carry assortments of colored paper, colored acetate overlays, felt-tip markers, and complete libraries of rubdown transfer type. (However, if you haven't had much experience using transfer type, take special care to remember legibility requirements when selecting a typeface and letter size. It will also be worth the effort to experiment a little to achieve proper letter spacing before beginning your art in earnest.)

Once your slide art has been prepared, you're ready to shoot it. We mentioned earlier that good registration requires a good copystand. The complexity of your art will determine the sophistication required in a copystand. Until recently, if you wanted highly accurate registration, it meant having your art shot on a large, expensive animation stand, such as those made by Oxberry, Forox, Sickles, or other manufacturers. (Normal 35 mm cameras — even expensive ones — do not provide sufficiently precise film transport and alignment.) Recently, however, more compact and less expensive systems that we know of have been introduced by Oxberry Division of Richmark Camera Service and by Maximilian Kerr Associates. These systems are specifically designed for producing accurately registered slides for multi-image presentations, and are even capable of producing such special effects as burn-ins, multiple exposures, and precise derivations.

If you prefer to have all of your slide art shot by a lab or specialty house that has a sophisticated, full-size animation stand, you'll need to sit down with the lab operator or representative and explain just how you want each piece of art shot. In addition, you should make notes on each piece of art relating to the desired field size and any special instructions pertaining to progressive disclosures or multiple shots from that piece of art. It's useful to remember that on these complex animation stands your art can be moved east and west (left and right), north and south (up and down), and even rotated on any point within the frame area, thus enabling you to achieve a form of screen action — or animation — by shooting several different slides from a single piece of art.

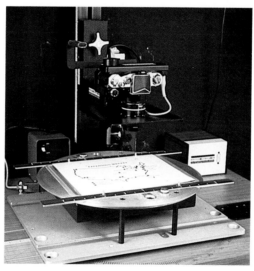

Copystands make it relatively easy to produce accurately registered slides and simplify the process for creating burn-ins, multiple exposures and photo derivations.

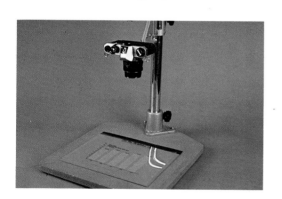

Pin-registered field guides are used to size and frame artwork to be shot on a copystand.

Field Guides And Registration

One of the best aids available for proper sizing, framing, and registration of artwork is a set of field guides. Because they are pin-registered, field guides provide accurate alignment between the artwork as it is prepared and the positioning of that artwork on a copystand. These guides also help make certain that the important part of the art will fall safely within the image area on the slide.

Also, because the artwork is pin-registered, you can add elements to the basic art by means of clear acetate overlays (called "cels") and still maintain accurate registration. This is particularly useful when your art is going to appear as a progressive disclosure on the screen (that is, when you're going to build up several elements, one on top of the other, or add a series of words to a list).

Accurate registration is also important when you're planning to do animation sequences in your multi-image presentation. By using cels and precise movement of your art on one of the more sophisticated copystands, you'll be able to make an object appear to move across the screen. Without the accurate alignment made possible by pin-registered art and a good copystand, this type of screen effect would require tedious measurements which—at best—could only come close to the precision needed.

- *The art*

- *The art*
- *How it was shot*

- *The art*
- *How it was shot*
- *The slides*

Progressive disclosures: (left) A series of type slides created using overlays. The first set of words is mounted on the art board: the remaining phrases are contained on three accurately registered cels. (above) Artwork created using the same process.

- *The art*
- *How it was shot*
- *The slides*
- *Programming*

The same care that went into writing a script should go into recording it.

The Message Of Sound

Your sound track can carry three types of audio information:

Words, whether a written narration or the quotes of interviewed authorities.

Sound effects, the aural details of a place, object, or event. Sound effects add another dimension to the presentation of a subject; they take you "on location." The sound of a whippoorwill singing, for instance, conveys as much information about the bird as do pictures.

Music, a more subtle, affective form of communication, is used to influence a listener's mood, usually by emphasizing visual action. Music can also be used to maintain a listener's attention during breaks in narration. In effect, music's message during these breaks is: "There is no additional narrative information at this time; please pay greater attention to the visuals."

When narration is backed by the impact of sound effects and music, sound becomes a powerful instrument of persuasion. Given the power of sound, you should strive to produce a professional sound track. Initially this means attending to your sound recording equipment and techniques. (In the section entitled Producing The Sound Track, page 156, you'll learn how to convert the individual segments of your recorded sound into a fully mixed sound track.)

144

Recording Equipment

The quality of your sound recordings will only be as good as your equipment. So if you want professional-level sound, you should plan to use professional-level equipment.

The two most important pieces of equipment you'll need are a tape recorder and a microphone.

You can choose among three different types of tape recorders—cassette recorders, portable reel-to-reel recorders, and studio recorders. Each type has advantages and disadvantages that you must consider when making a choice.

Cassette recorders aren't practical for recording sound to be used in a multi-image presentation. This limitation isn't a factor of their quality; many of the cassette recorders on the market today record and play back sound with the fidelity of reel-to-reel recorders. Rather, it's a factor of a cassette recorder's smaller tape size (about ⅛ inch [3.2 mm]) and slower recording speed (1⅞ inches per second) that makes it incompatible with the ¼-inch (6.4 mm) tape used on fast-speed (7½ ips to 15 ips) reel-to-reel tape recorders used by sound editors. For an editor to use material originally recorded on a cassette, he or she would have to rerecord (dub) it onto ¼-inch tape. The result, even when using the best equipment, is usually some loss of sound quality.

Of course, cassette recorders are highly practical if you're recording interviews as part of your preliminary research. Their compact size and ease of operation make them ideal for situations where the emphasis is on gathering information.

Portable reel-to-reel recorders of professional quality can be used for almost any sound recording assignment. If necessary, they even can be used to record sound in a studio.

With tape recorders, as with almost all other products, you usually get what you pay for. So, in general, the more you spend for a portable reel-to-reel recorder, the more sophistication you'll receive. This is particularly true with such features as the tape transport mechanism, the preamplification components, options used for recording lip-synchronized film sequences, etc.

Most portable tape recorders use five-inch and seven-inch reels of tape—enough for 20 to 45 minutes of taping at a speed of 7½ ips.

Sound With Film

If you're planning to shoot film using on-location sound, there's an additional factor to consider when selecting a recorder: Will you shoot your footage using single-system or double-system techniques?

In single-system filming, the camera serves a dual role. It exposes the film and simultaneously records sound on a magnetic stripe that runs along one edge of the film. It also eliminates the need for special features on your tape recorder, such as crystal sync. But single-system sound, while easier to record, has its drawbacks. Its quality usually is not as good as that recorded using double-system techniques (separate camera and tape recorder). And the sound for 16 mm motion pictures is displaced 28 frames from the corresponding frame of action—the distance between the lens aperture and the sound recording head in most professional cameras. In double-system filming, you use both a camera and a reel-to-reel tape recorder. Both units are linked, using either sync-pulse or crystal-sync equipment, to assure synchronization between film speed and tape speed. You can find more information about the equipment and techniques for double-system recording in books dealing with sound recording for film.

Studio tape recorders are highly sophisticated and extremely costly. Unless you're equipping a sound studio, you won't be in the market for this type of equipment, which is best used to full advantage by a professional sound-recording engineer.

The factors that make studio equipment so expensive are its reliability and durability, as well as a far superior signal-to-noise ratio, low distortion, and wide range of capabilities. For example, the output from numerous microphones can be fed into these units, allowing the recording system to tape a single narrator or a 200-member choral group. In addition, these tape recorders have excellent signal-to-noise ratios and low distortion, and they're designed for more reliable and durable operation. Instead of adjusting the recorder with a few knobs and dials on the unit itself, the sound engineer sits at a mixing console, where he or she controls dozens of equipment functions through an array of switches, "pots," and meters.

What equipment is best for you? The answer to that question depends on two factors: First, how much money can you afford to spend on recording equipment? And second, how much use will you get out of the equipment you buy?

In general, you should buy the best possible recorder offering the features you're most likely to use. Don't buy features or options you won't use. On the other hand, don't cut corners on the basic unit itself.

You must sort out your specific requirements. But at the very minimum, make sure the equipment you buy contains the following features:

- reel-to-reel recording
- ¼-inch tape capability
- ability to record on full track for best signal-to-noise ratio (although quarter-track and half-track recording are also acceptable)
- a recording speed of at least 7½ ips

A recorder with these features will serve almost all of your needs. (See the section on Sound Equipment, page 182, for more information on using a tape playback unit during the presentation.)

Choosing A Microphone

Selecting a high-quality microphone (or microphones) is as important as the selection of a recorder itself. You can own the most sophisticated tape recorder on the market today, but if you plug an inexpensive microphone into it, you're going to end up with poor sound.

Microphones are categorized according to their sound acceptance patterns:

Omnidirectional microphones pick up sound in an arc sweeping about 300 degrees across the face of the microphone. This is the type of microphone sold with most inexpensive tape recorders. If you've ever used one of these recorders, you know the microphone picks up a great deal of ambient sound. For this reason, it's not always the most practical microphone to use for professional sound.

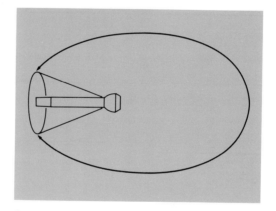

Recording pattern for omnidirectional microphone (right).

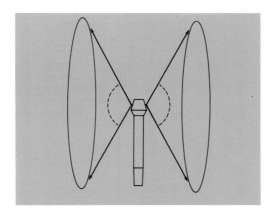

Recording pattern for bidirectional microphone (left).

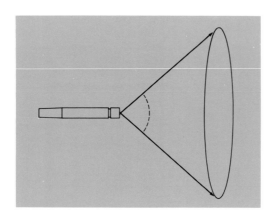

Recording pattern for cardioid microphone (left).

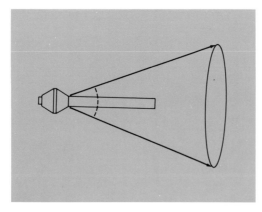

Recording pattern for ultradirectional microphone (left).

Bidirectional microphones are used mainly for one-to-one interview situations. They're sensitive to sound at both the front and back of the microphone and are impractical in most other situations.

Cardioid microphones are the most practical mikes when your sound source is contained in an arc of about 80 degrees in front of the mike. This pickup pattern allows you to focus on the principal source of sound while also recording some ambient sound to recreate the atmosphere of the location.

Ultradirectional (shotgun) microphones record sound in an arc of about 40 degrees in front of the microphone. This makes them extremely useful when you want to record isolated sounds, such as might be the case if you were trying to record conversation on a city street.

Microphones also can be classified according to the type of use they get.

General-purpose microphones, the type most frequently used with portable tape recorders, are designed to absorb the jarring and rough handling often connected with traveling and on-location use. To build in durability for this kind of application, manufacturers have had to sacrifice some sound quality. But not much. In fact, it would be hard for an untrained ear to pick up the difference in sound quality between narration recorded using a portable mike and the same material recorded using a studio mike.

Studio microphones are extremely sensitive, picking up sound—or frequency response—beyond the recording capability of most portable equipment. Unless you're working in a sound recording studio, you'll have little call for microphones in this category.

Specialized microphones. If you're planning to film a lip-sync interview, you might want to use one of two specialized microphones. One is the lavaliere microphone. This is the type of microphone you often see clipped to the tie, lapel, or blouse of a television newscaster. Although miniature in comparison with other microphones, the lavaliere mike does an adequate job in picking up a speaker's voice. Its major disadvantages are the need to run wires from the mike to the recorder, the blockage of sound sometimes caused by the speaker's chin, and the tendency of the microphone to pick up scratching sounds caused by the rustling of clothes.

The second specialized microphone is the wireless or radio mike. This type of mike uses a miniature VHF transmitter to broadcast sound to a receiver that's linked to a tape recorder. These microphones are equal in sound quality to the lavaliere mike, while being less intrusive. But some of the older ones have a major drawback: The receiver — and thus the audiotape — may pick up unwanted sounds such as FM radio signals and chatter from citizen's band radios.

With all these microphones to choose from, which one is best for you? That's hard to say, but two suggestions might help you narrow your choice.

If you're planning to use your recorder and microphone solely for multi-image productions, you'll probably be better off with a cardioid microphone. This microphone allows you to record sound without having to be unduly concerned about sound originating from behind the microphone. (This doesn't mean people can carry on a conversation outside the range of the mike; it just means that the normal activity of your sound crew or film crew is less likely to be picked up by the mike.)

If you're going to use your microphone for other than multi-image production, such as for recording meetings, conferences, and other events, you're better off buying a good omnidirectional microphone.

The other types of microphones are more specialized, so unless your organization also maintains a sound studio, you'll probably have little need for them. And when you do need them, you're probably better off renting them or, more likely, hiring a sound specialist who can supply both the equipment and the know-how.

A wireless microphone and radio transmitter.

149

Recording Sound

Although this is, by now, a familiar message, it must be said again. Sound recording, like scriptwriting and photography, is a specialized field, one requiring considerable study and on-the-job experience to master. This book is not meant to provide that experience. There are a number of books available covering sound-recording techniques. If you want to master the techniques of sound recording, they are the place to start.

A professional sound recording studio offers the multi-image producer a number of specialized services.

The suggestions you'll find in the remainder of this chapter presume some knowledge of sound-recording practices. Their purpose is to help you apply your experience with sound to the demands of a multi-image presentation.

You'll record sound in one of two places: a sound studio or "on location." Each situation places slightly different requirements on you.

1. Working in a studio. A sound studio is a room designed and constructed specifically for recording and mixing sound. When constructed properly, the studio is isolated from external sounds that might interfere with recording. The recording area itself should be acoustically balanced, neither too flat nor too tinny and sharp, with little echo or reverberation present.

You'll use a sound studio to record the narration for your presentation (with the exception of segments recorded on location), to record music and library sound effects, and to mix the various elements into a final sound track.

If you don't have access to a sound studio within your organization, you'll have to turn to outside sources. Most commercial sound studios sell their time by the hour. To determine the amount of time you'll need to **record** your script, read the script aloud, keeping track of the time with a stopwatch. Don't rush through the script and don't try to slow your delivery unnaturally. Just read it as if you were speaking to an audience, at a rate of about 125 words a minute.

When you have your reading time, double that figure. That's about the time you'll need in the studio. If you're using an inexperienced narrator, allow additional time for orientation and a higher-than-average number of retakes.

Picking A Studio

Finding a studio to work with isn't difficult—if you know what to look for.

First of all, you'll want to judge the experience of the people operating the studio. Are they professionals? Have they recorded sound for multi-image presentations (and have they mixed multi-image tracks, something you'll want them to do at a later date)? Or has most of their experience been gained recording radio commercials or musicians? Of course, this sort of experience doesn't mean the people aren't competent, fully capable of recording and producing sound tracks. What it does mean is that they'll probably have to do some learning themselves as work progresses. If you can allow time for learning—and if the studio personnel seem willing to learn—you can make an exception to the "experience counts" rule. But all things considered, you'll probably be further ahead working with a studio that has had multi-image experience.

After the experience of the studio personnel comes the quality of the studio itself. This may be difficult for you to judge, especially if you have little or no experience in sound recording and mixing. But you must try.

Start with the sound-isolating properties of the studio. Do you hear external sounds filtering into the studio? If you can hear horns honking, dogs barking, someone walking down the hall, or air conditioners running, so will the studio's sensitive microphones.

Next, judge the acoustics of the studio proper. As you talk, listen to the tones of the voices in the room. Are they dull, having been absorbed by the room's acoustics? Or are they sharp and tinny? Do you hear echoes or reverberations? Or are the sounds clear and true? Naturally, it's the latter qualities you're looking for.

The experience of the sound-recording engineer and the condition of studio facilities and equipment are key considerations in choosing a sound studio.

Examine the recording, mixing, and control equipment and the control room itself. Does the equipment look relatively up to date and well maintained? And is the control room neat and well organized? Or is it littered with strips of tape and pages from old recording logs? In general, pick a well-organized, well-maintained studio to do your work.

You can make this judgment easier if you visit at least three recording studios. These visits will not only increase your understanding of the recording process, they will also allow you to make comparative judgments of the studios.

151

When you're recording narration at a studio, the sound engineer will assume responsibility for the quality of the recording. Your job at this time is to direct the narrator. This means attending to such factors as:

Delivery. Is the script to be read with a solemn, serious delivery? A lighthearted delivery? A straightforward, matter-of-fact, businesslike delivery? The urgent delivery of a newscaster? You must make that decision, then communicate it to your narrator. Because discussions of delivery usually involve subjective interpretations, the narrator, especially if he or she is a professional, will want to rehearse several times to find and refine the proper delivery. By all means take advantage of these practice runs. (Indeed, if the narrator doesn't suggest a few trial readings, you make the suggestion. It's better to refine the delivery before recording begins, rather than after, when repeated starts and stops threaten the overall quality of the recording.)

Professional narrators look for suggestions, coaching and encouragement from a director.

Pace. Is the pace to be quick and urgent, suggesting matters of immediate concern? Is it to be more leisurely and casual, suggesting friends discussing a matter of common concern? Or is it to be crisp and businesslike, suggesting matters that are to be dealt with in a matter-of-fact manner? Once again, you make that decision, then communicate it to your narrator.

During the recording session, your ear is the standard against which to judge the narrator's delivery. You know what "sound" you want in the narration; so if the narrator isn't producing that sound, stop the recording session and let him or her know. But don't limit your comments to criticism; offer suggestions for improving the delivery to meet your standards.

And don't be hesitant about offering suggestions or constructive criticism. Narrators **know** they're working for the director. They **expect** the director to comment on their delivery. Narrators aren't mind readers—and they don't want to be treated as if they were. Taking direction is part of their profession, so you'll be working with them—not against them—when you offer it.

Music for your sound track can be recorded in one of two ways. If you're recording an original composition, the experience of your recording engineer becomes critical. Recording music involves greater problems and considerations than recording voices. So the studio you hire for this assignment should have experience in mixing and recording musical arrangements. (Music played from a published composition [sheet music] is copyrighted and requires that you obtain permission to use it. You may also have to pay a royalty fee.)

An easier way to record music, of course, is from a record produced and legally cleared for public use. This approach avoids problems arising from possible infringement of copyright laws. As mentioned earlier, this "library" music is paid for by the number of "needle drops" you make in recording your musical track. The drop fee is set by the library. Once you pay the fee, no other clearances or permissions are required from the library owner.

If, on the other hand, you want to use music from an album or record **produced for retail sale,** you must ask the record company, the recording artist and the publishing company for permission to use the music. They will probably want to know how you plan to use the music and for what type of presentation. If they agree to your planned use of the music, they will give you written permission. **Do not use the music without written permission from the copyright owner.** On occasions, producers have received oral permission from a representative of a record company to use a particular selection of music, only to learn later that the person who gave permission had no authority to do so.

When you use music from a record produced for retail sale, you usually must pay either a royalty or flat fee to the recording artist every time the music is played. Naturally, this is much more expensive than the flat fee (one-time needle drop) paid to the owners of library music. So if you plan to use music from a popular album, make sure the selection you want is worth the time and money you'll have to spend.

2. Working "On Location." If you're recording sound on location instead of in a studio, you have additional problems to deal with, foremost of which is the problem of ambient sound. Usually, however, if you use an appropriate microphone and a little common sense, you can minimize the problem. Two suggestions will help make your recording job easier.

Auditioning A Narrator

Unless you've worked with—or at least have heard—your potential narrators, you're going to have to select a voice through the use of audition tapes. These tapes demonstrate the style, tone, and range of the narrator's voice. In smaller cities, you can get copies of these audition tapes from the narrators or from studios that keep audition tapes for just this purpose. In larger cities, you may have to work through a narrator's agent.

If you're hiring only a narrator's "voice," an audition tape provides enough information for you to make a judgment. But if you're planning to use your narrator on film or as a live personality in front of an audience, then appearance becomes a factor. You'll have to see as well as hear the person.

Here again the criterion for judging is a simple one: Does the narrator's appearance match the mood you're trying to set? If you're producing a presentation for a young, exciting marketing organization, you might not want an older-looking narrator for a spokesperson. On the other hand, if you wanted to convey a sense of mature judgment and decision-making ability, an older-looking narrator might present the right appearance. Your own intentions must serve as a guide.

When choosing a narrator to appear in your presentation, you also should consider the person's ability to work in front of a camera or a group. Many narrators spend their entire careers working behind a microphone and are no more comfortable in front of an audience or camera than the average person. So, look for some film or podium experience if your narrator is to appear in your presentation.

153

• Don't record your narration in a normally noisy location. Although it's possible to record a narrator riding a merry-go-round that's circling to the blaring sound of calliope music, it just isn't worth the effort. Unless your intention is to prove the sophistication of your sound-recording techniques and equipment, don't go looking for problems to solve.

• Don't overlook obvious sources of unwanted sound. And when you find them, eliminate them. Before you begin a recording session, look around your site for people, objects, or events that are likely to ruin a take. If you're recording in an office, for example, beware of telephones. Have a secretary, receptionist, or operator intercept all calls headed for your recording site. If you're recording sound at someone's house, make sure the family watchdog is tied up in the backyard, where its barking won't interfere with recording. You get the idea: If you eliminate possible interruptions ahead of time, you won't be bothered by them later.

Recording sound "on location."

The best equipment to use when recording narration on location is a portable reel-to-reel recorder such as one of those mentioned earlier in this chapter. The best type of microphone to use is a cardioid microphone if you want to pick up some ambient sound from the location; use a shotgun mike if you want to eliminate as much peripheral noise as possible.

Location recording usually involves working with three sources of sound — spoken words, sound effects, and wild sound.

The words will be spoken either by your narrator or by the subject of an interview. Working with a narrator on location is no different than working with one in a studio. You have to supply instruction, direction, and encouragement; then let the narrator take it from there.

Recording an interview, however, presents a different set of problems to solve. First you have to judge the suitability of the person you want to interview. Then you must prepare **yourself** to ask the questions and to direct the recording session.

Choosing the right person for an interview requires editorial judgment on your part. You have to judge whether or not the person is worth recording. If your subject's words merely repeat a commonplace fact or opinion, perhaps you would be better off to have your narrator quote or paraphrase the person. If, on the other hand, your subject offers new, controversial, highly opinionated, or emphatically spoken information, you'll probably want to record his or her words.

Preparing yourself for an interview means thinking out the order and phrasing of your questions. Unless you have hours of tape or film to devote to a single interview, the worst interviewing technique you can use is to ask a broad, open-ended question. This leads to a rambling answer — and an ineffective sound track.

Interviewing Before The Interview

How do you know if a person's words are worth recording?

Just ask.

Following a fairly common practice, you can interview potential on-tape or on-camera subjects early in your research, recording the questions and answers on a cassette recorder. (Many of these interviews can be conducted over the telephone, to save the time and expense of traveling to many locations.) This research is then used to write a script — and to select those individuals whose facts, ideas, or opinions vividly and emphatically support the theme of a presentation.

Knowing what these people will say — and where in the presentation it will be used — you can call the subjects again, asking if they would mind answering a few more questions, this time on film or audiotape. In this way, only the most suitable people are interviewed and only the most pertinent questions are asked and answered. This not only saves film and audiotape, it also produces a more organized sound track.

The more practical approach is to elicit information using a number of specific questions. This allows your source to cover material in shorter segments, and that usually means a better organized and more emphatic answer. If you're filming the interview, this approach also allows you to move your camera to different positions. This reframing of the scene gives your film editor the footage he or she needs to avoid "jump cuts." (Jump cuts occur when two widely separated frames from one sequence are joined. This abrupt transition causes the image of the person or object in the film to "jump" disconcertingly as it passes the gate in the motion picture projector.)

Again, if you're filming the interview, your preparation must also include plans for directing the sequence. Tell your subject where to look. If you want your subject to pick up or refer to some object, explain how you want it done. If your subject is relaxed and confident enough to move during the filming, explain what action you would like performed.

Your directions — whether for filming or taping — should also include instructions on how to answer a question. This doesn't mean coaching; it means establishing the context for the subject's remarks. In normal conversation, a person answering a question picks up from where the questioner left off, with both responses forming the framework for the discussion. In a multi-image sound track, however, interviews are generally edited for brevity and emphasis, so the initial question is cut out. If the person doesn't restate the question as part of the answer, the reply may seem incomplete.

An example will make this point more clear. If you asked a subject whether or not he or she agreed with a certain technological forecast and the reply was, "Of course I do, for the following reasons . . . ," you would have your answer, but with half of its framework missing. Had the subject restated the question with his or her answer, "I think the technological forecast is essentially correct because . . . ," the reply would be fully framed.

The first step in the production of a sound track is usually the editing of the narration into a rough "scratch track."

Producing The Sound Track

Synesthesia. It's a great word to consider when you talk about the importance of sound to a multi-image presentation. Its dictionary definition is "a concomitant sensation." In terms of everyday experience it means seeing colors when listening to sounds or smelling coffee when you see a vivid, full-color advertisement focusing on a steaming mug of coffee. In short, one sensation produces effects in the realm of another sense.

The effects of synesthesia are important to producers creating multi-image sound tracks. Producers often "hear" the music and effects they want when they look at their visual sequences. And many producers feel more comfortable when they have a sound track against which to plot their visual sequences. In a very real sense, the music and effects help them "see" how a sequence should be developed.

So when the effects of synesthesia grip you, go with the flow. Let sound and visuals influence each other. One way multi-image producers do this is to create a rough sound track — or "scratch track" — to use as the basis for visual planning and programming. They edit their scratch track until it sounds right for the show. Then they select and program their visuals as if creating the choreography for a dance. Once the union of sound and visuals feels right, the final sound track is mixed.

There are three major advantages to producing a scratch track before selecting a sequence's visuals.

1. The sound track will help you determine the mood or feeling of a sequence, or of the entire presentation. Naturally, the visual style and programming should be in harmony with this mood. For example, if you're using an upbeat, brassy track, you'll probably want to use bright, vibrant

visuals and a lot of quick cuts and flashing. On the other hand, if you're using a slower, more romantic track, you'll probably want to use slides with soft colors and soft focus, in a sequence programmed with a number of slow dissolves and wipes.

2. The music used in your sound track influences the pacing of your presentation, which in turn influences the number of visuals you're going to need. So your scratch track will help you pinpoint the number of visuals you need to complete a sequence or an entire presentation.

3. Producing a scratch track is particularly useful if a sound track is to be an involved, multi-track mix. A complex 10-track mix—with multiple voices, music, and a variety of sound effects—is difficult and costly to change just to add or delete a few seconds from a visual sequence.

An exception to this practice of developing your sound track early involves presentations using motion picture sequences. Because of the time it takes to get work prints, edit and conform the film, and send the negative out for answer prints, it's a good practice to get this portion of your production under way as early as possible. (These activities will be discussed shortly.) If, on the other hand, the length of your film sequences will be directly related to the timing of your sound track, you'll have even more reason to work on the sound track first—and early.

And to do this, you're going to need some equipment—or the services of a good sound studio. If your organization has the people and facilities, you're already prepared. If not, you may want to consider investing in some equipment—if your sound track isn't going to be too involved—or hiring a recording studio that can handle your more complex needs.

Preparing A Recording Script

An experienced narrator knows how to give a script the inflections and pacing necessary to make the narrative interesting and effective. You can make this job easier by using some professional "tricks" and methods when preparing a recording script.

To begin with, provide the narrator with a clean double-spaced typed copy of the script—preferably in advance of the recording session. Don't clutter it with unnecessary parenthetical remarks or descriptions of visuals. The best practice is to have the narration typed on the right-hand half of the page. This serves two functions. First, it leaves you space on the left to put visual descriptions (if necessary), and secondly, it gives the narrator shorter lines to read. No matter how experienced your narrator may be, it's a simple fact that narrower columns of copy are easier to read than wide ones.

If there are to be pauses in the narration, indicate these on the script. For instance, if your narrator must read a list of items that must be paced to match slide changes, indicate the "pause points" on the script and point them out before the recording session begins. It saves time and will sound more natural if the narrator builds these pauses into the delivery, rather than having an editor splice in short sections of room tone later.

Finally, don't make it necessary for your narrator to read sentences that carry over from the bottom of one page to the top of the next. Make your page breaks coincide with natural breaks in the narration. This will make the script easier to narrate and decrease the possibility that a page turn will be audible behind the narrator's voice.

The Role Of Sound

For a moment, let's talk about the role of sound in relation to a multi-image presentation. It's vitally important. If you've ever watched a television special and lost the sound portion, you realize that much of the effect is gone. You're dealing with an **audiovisual** medium, where the **audio** portion is often crucial. So you're only kidding yourself if you attempt to throw together a sound track without the necessary expertise and equipment. To illustrate this point, let's consider a multi-image presentation conceived to introduce a new product line to a national dealer organization. And let's say it'll use 15 slide projectors and an electronic computer programmer. Further, let's say the budget for original photography will approach $15,000. With these considerations, it doesn't make much sense to put together a sound track on a quarter-track stereo tape recorder designed for home use.

So keeping in mind the level of expertise and the sophistication of equipment needed for the kind of sound track your show will require, let's look at some basics. Earlier, we talked about recording sound—in a studio and on location. The equipment we discussed was high-quality, professional gear. When it comes to editing your tapes and mixing a sound

track, you should consider no less. Good professional quarter-inch equipment is available with full-track, half-track, or quarter-track capability. That means the audio signal is laid down on what is nominally the full width of the tape, half the width, or one quarter of the width respectively. Moving up in sophistication (and cost), half-inch equipment is also available with full-, half-, or quarter-track capability. Equipment beyond these capabilities is generally required only by large recording studios and must be used almost constantly to pay for itself. Included in this type of sophisticated equipment are the one- and two-inch units that can have up to eight tracks (one-inch [25 mm]) and as many as 24 or even 32 tracks (two inch).

What equipment and facilities will you need to produce a multi-image sound track?

If you want to handle the recording for your presentation, you'll need a studio. The studio should include a soundproof announcer's booth and a control room, with an insulated window between the two. For the booth, you'll need one or more of the high-quality microphones described earlier and an intercom system to communicate with your narrator. In the control room you'll need one or two good two-track tape decks and at least one four-track deck to permit mixing. Then you'll need a turntable with a small preamplifier, an audio mixer, an equalizer, a couple of good power amplifiers, two loudspeakers, and maybe a reverberation unit. This casual listing of what it takes to set up a recording capability isn't meant to intimidate you; it merely emphasizes the fact that it's not the sort of project you should begin without serious thought. Many production groups have concluded they need this type of equipment and have invested the money required. If this is the route you choose, then your first step should be to contact a good audio consultant to help you design the facilities and recommend specific pieces of hardware to meet your particular needs.

Now let's get on to the subject of producing your sound track. If you've never been in a recording studio during a taping, the first point you should understand is that you don't have to complete the sound track in one pass. That also holds true for the narration. Instead, you'll follow a process that typically goes like this:

Once you've gone over the script with your narrator and have let him or her know what style of delivery you want, your sound engineer will get a voice level—that is, your narrator will read some of the script while the engineer adjusts the input level to the master tape deck so the VU meter peaks at "0." After that, the recording session is ready to roll.

Misreadings are a natural part of a recording session, so when your narrator misreads or mispronounces a word, mark the script and let him or her repeat the misspoken sentence. Then keep going. When the whole script has been recorded, ask your narrator to sit quietly in front of the microphone while the equipment records about one minute of "room tone"—the ambient sound present in the recording booth. (Recorded room tone is used to add time between occasional sentences or paragraphs when an audiotape is edited. Using unrecorded tape for this purpose will produce a distinct difference in sound when you play back the narration.)

To edit the mistakes from a tape, an editor runs through the tape until he or she reaches those spots marked on your copy of the script. Then the editor marks the tape at the mistake and at the point where the narrator began the next good "take." Using a splicing block and a sharp razor blade, the editor cuts out the bad section, then splices the good ends together. In this fashion, the editor removes all the "goofs" and makes adjustments such as opening up or tightening space

between paragraphs, sentences, or even words. Occasionally, an editor may find it necessary to edit out the sounds of pages turning or loud breaths. When this has to be done, be sure the editor replaces the removed section with an equal amount of recorded room tone. This maintains proper timing.

Once this is done, you'll have your master voice track. However, because it is the only thing on the tape, you'll be able to come back to it later and adjust the pace, should that be necessary.

The editing process is similar when part of your voice track is to come from interviews taped on location. When you want to use only particular parts of an interview, you may find the editing job easier if you have the whole interview transcribed. Then pick out the usable sections from the transcription. This will enable your editor to find these parts on the tape more quickly.

One caution: The "magic of editing" allows you to excerpt words, sentences, and whole paragraphs from an interview, and those alterations can be undetectable. That's where—and why—you must be careful. If in the course of shortening an interview you change the intent of what a person said, you could be treading on thin ice. You are, in effect, "quoting out of context." Even if your subject knows beforehand that you'll be editing his or her remarks, you still have an ethical responsibility to maintain the integrity of what was said.

In addition, make sure you listen to the audiotape before "paper editing" the transcript. You may link passages or words that look natural on paper but sound choppy on audiotape.

Using Music And Sound Effects

Although your narrator's voice may form the core of your sound track, music and sound effects animate that core and give it color. This makes the selection and use of the right music and sound effects extremely important.

People starting their first multi-image production often ask when to use music and sound effects. Unfortunately, there's no set formula to follow. Some presentations work best when music and effects are present throughout the whole show, coming up and receding between narrated segments. By contrast, we've all seen presentations with continuous musical backgrounds that bring to mind the monotonous sound tracks from training films of the Fifties. Still other presentations work best when music and effects are restricted to opening and closing sequences, where they act as punctuation to set the tone, or function like an exclamation point. Unfortunately, this approach is too often used as a way out when, in fact, the show cries out for more "color."

Each show is unique in this sense, and the type and amount of music and effects used must be determined on an individual basis, depending upon the audience, the objective, the tone of the message, and even the location and environment in which the presentation will be shown. The only suggestion we'd be inclined to offer is this: As colorless as a presentation might seem without sufficient music and effects, it won't be as bad as a presentation that is punctuated with unnecessary, meaningless, and distracting music and effects.

As we mentioned earlier, having an original musical score composed for a multi-image presentation is a luxury—something you must be prepared to pay for. The more common practice is to use library music; but it seldom seems to fit the needs of the presentation "straight off the disc." The piece is either too long or too short, or it builds when you'd like it to recede, or recedes when you want a crescendo. This is where the experience of your sound engineer or editor comes into play. And this is where a good multi-track recorder will justify its cost.

By using separate tracks for music segments and sound effects, a sound mixer can control their respective levels in relation to the voice track. In many cases, this alone overcomes the fact that the music doesn't quite "fit" the way it might if it were scored for the presentation. But more often than not, even this degree of flexibility won't solve the problem completely. So the next step is editing the music. Unless your sound editor has a good ear for music, this could be an arduous—if not impossible—task. This, of course, is why we suggested earlier that you pick a good sound studio to do your work if you don't have in-house capability. But as complex as this process may appear on the surface, a good sound engineer can lengthen or shorten a piece of music with relative ease. And because it's a talent that is either innate or developed through much practice, we won't attempt to analyze—or teach—it. We will, however, emphasize the fact that it can be done—by deleting whole measures or looping others—and **is** done as a matter of practice when a piece of music doesn't fit the way it should. So when you're ready to add music or effects to your voice track, and can't afford to have it scored to fit, don't feel limited to the use of "as is" library music. Feel free to tell your sound engineer what parts you like and what parts you want edited. If your requests are reasonable, it can be done, and you'll end up with a track that's every bit as good as one scored to fit your show.

Once you've laid down all music sequences and sound effects with your voice track, you're ready for the final mix. Earlier we mentioned the use of a scratch track for preliminary timing with your visuals. If there's any chance you may want to alter parts of the sound track, now is the time to make those changes—with the scratch track. Your engineer can produce a scratch track by duping the multi-track master onto a single-track copy. Use this copy while you're developing your visual sequences.

Once you've locked in your timing, you can come back with any changes and do the final mix. You can plot that final mix either by means of notations in the margin of your script or on a separate mix-down chart, if the mix is to be relatively involved. The most effective method is to plot the changes in music or effects against a time frame, as is illustrated in the photograph at the right. With various pieces of music and sound effects on separate tracks, you'll be able to adjust levels, change fade points and, in effect, create whatever total impression you want your final sound track to convey.

Above all, keep in mind the fact that you—as the producer—must represent the client in judging what sounds right. You will be the one responsible for the final outcome. So make certain your final mix is exactly what you want. A good sound engineer can be very creative and can offer many useful suggestions that might improve the final result. On the other hand, his or her greatest service to you is as a technician. You're familiar with the objectives of your multi-image presentation. You know what kind of effect that sound track has to have on your audience. So you're the one who must direct the final mix. And if you aren't satisfied with the results of the first attempt at a mix, don't hesitate to tell your engineer what you want changed. It isn't uncommon to make two or three passes at a final mix before the right one is achieved. When the narration has just the right pacing, and the music comes up or goes under at just the right points and at just the right levels, and the sound effects hit at just the right spots, you'll all know that you've got it.

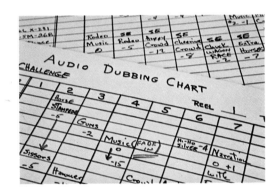

Peter Chermayeff

Peter Chermayeff:
Creating An Experience
With Sound

It's one thing to capture the essence of a city on slides, still another to duplicate its ambience on audiotape. When you do both, you have a multi-image presentation that becomes as celebrated as the city it portrays.

You have **Where's Boston?**

Where's Boston? is a 55-minute grand tour of the Boston area. But it's also more than a tour. In its overall conception, **Where's Boston?** is a multi-image, multi-media study of the history. people, customs, culture and **sounds** that define and distinguish the city's many communities.

"People forget they're watching a multi-image presentation and become part of an experience," says Peter Chermayeff, a partner in Cambridge Seven Associates and executive producer of **Where's Boston?**

Where's Boston? was created as a Bicentennial attraction for the city, under the sponsorship of the Prudential Insurance Company of America. Responsibility for production of the show and a presentation pavilion was given to Cambridge Seven Associates, a Boston-based firm of architects and designers. (The firm's credits include the interior architecture and exhibits for the U.S. Pavilion at Expo '67 in Montreal and the New England Aquarium in Boston.) Although originally conceived as a presentation whose life would coincide with the city's Bicentennial celebration, **Where's Boston?** demonstrated it had more enduring appeal. It has since become a permanent attraction in the city's historic Faneuil Hall-Quincy Market area.

From the beginning of the project, Chermayeff knew what he wanted to achieve: "I wanted a portrait of the city in documentary style, a presentation with convincing realism and emotional power, the city speaking for itself with no narration, no actors, no scripted interpretation."

To accomplish this goal, Chermayeff concentrated on the presentation's design: it would have a deliberately non-linear structure, made up of almost randomly organized segments and fragmented impressions. Further, his concept called for the use of all slides, no motion picture film, and for the programming of the presentation to be sophisticated but not flashy.

To bring the production to life, Chermayeff hired Rusty Russell, designer and director of **The San Francisco Experience** and **The New York Experience,** to create the detailed design of the show and to direct overall production. They also contracted for the services of Dimension Sound Studios of Boston and its chief engineer, Thom Foley.

The early involvement of a sound studio was necessary, says Chermayeff, because "the content of the sound track would be the primary design element." The detailed visual design, editing and programming were done by Russell, who worked against the finished track.

To create the voice track of the presentation, Russell spent almost nine months interviewing citizens of Boston. His recordings were edited, catalogued, transcribed and assembled into a 55-minute rough cut of the voice track as production moved along.

In addition to recording the voice track, Russell and his crew began recording sound effects throughout the city.

While interviews and sound effects were being recorded around Boston, Foley was recording the music for the presentation in the studio, using the timing of the voice track to create the timing for the musical segments. The musical theme for **Where's Boston?** is based on an 18th-century hymn written by a Boston tanner. An original composition based on the hymn was created by Richard O'Connor, a Boston composer, and performed by the Wind Ensemble of The New England Conservatory of Music. Variations of the theme were played by a night club jazz group, a Jamaican steel band, a fife and drum corps, and with individual instruments. Another variation of the theme was recorded by the Old North Choir.

"When we were through we had almost one hundred music numbers," said Foley. These 16-track recordings were then mixed down into separate, discrete quadraphonic music tracks.

Assembling the various elements of the total track – which were recorded on almost 72 miles of audiotape – required Russell and Foley to combine equal amounts of creative energy and reflective patience. They assembled the elements on 16 tracks four each for voices. sound effects. music. and the swing music that provided transitions between scenes The thousands of segments that would eventually make up the final track were transferred from the original ¼-inch and ½-inch audiotapes to 2-inch, 16-track tape – **one segment at a time**.

The process was a mixture of art and engineering. "We would test the placing and timing of an element and its relationship to the other elements before we'd make the transfer," says Foley. "Sometimes we'd adjust the positioning three. four or five times before we were satisfied." Working at that deliberate pace, it took Russell and Foley three months to assemble and position all the elements of the track.

Mixing the 16 tracks down to the four tracks used in the presentation also was a slow and deliberate process, requiring almost 100 hours of work over a 10-day period. "We worked in short segments," says Foley, "no more than one or two minutes in length. We'd test a segment maybe a dozen times, always keeping track of our levels, then when we were satisfied, we'd record the mix."

The completed production is shown on eight screens using 40 projectors controlled by an Arion 832 programmer. The sound track is played back on a Scully 280B tape recorder, linked to a dbx noise-reduction unit. The four individual tracks then go through a filter, where all information below 100 Hz is taken off and channeled through an amplifier to two bass-reinforcement speakers located beneath the screens in the front of the **Where's Boston?** theatre. The remaining sound is channeled through one-octave band graphic equalizers into Crown amplifiers, then to four JBL loudspeakers, one located in each corner of the room.

The effect, says Chermayeff, is "to create a sense of environment in the room."

It's a sound track that faithfully recreates the ambience and captures the spirit of Boston – an extraordinary accomplishment honoring an extraordinary city.

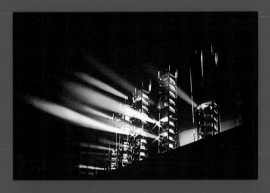

Forty projectors illuminate eight screen areas in **Where's Boston?** theatre.

Thom Foley, chief engineer of Dimension Sound Studios, Inc. (Photo: Dimension Sound Studios)

163

Editing Your Film Segments

When you developed the initial planning for your show, you were faced with several alternatives regarding motion picture segments. First, you had to choose between 16 mm and 35 mm as your production medium. (We briefly discussed other film sizes, but indicated that these two were the most commonly used.)

Second, you had to decide if your film segments would be silent, voice-over sound, sync sound, or a combination of two or all three. Once these film segments are shot, you have to combine the various takes into coherent sequences; then you have to integrate these sequences into the overall flow of your multi-image presentation. This is the task of film editing.

You can find many books that teach how to edit motion pictures. If you aren't familiar with the process, by all means study them. There are differences, however, between editing a motion picture and editing motion picture sequences to be used in a multi-image presentation so we'll review the whole process briefly, pointing out the similarities and differences.

For the sake of this discussion, let's make several assumptions:

• that your motion picture sequences were shot on 16 mm color negative film (the most frequently used film format for multi-image productions);

• that your cinematographer sent the film to a processing lab with a camera log and requested a work print (this is standard practice);

• that while the film was being processed, you had the sound track transferred from ¼-inch tape to 16 mm mag tape in preparation for editing the sync footage (again, standard practice).

When the work print arrives from the lab, the actual work of editing begins.

The first step is to get the 16 mm mag tape synced up with the footage. Since you probably have multiple takes on sync scenes, you'll want to sync up only the good takes. (This information comes from the camera log, on which the acceptability of each take should have been indicated at the time of the shooting.) After the film and sound are synced up, you should review all of the remaining footage, including silent footage shot for cutaways and close-ups.

Film Editing Hardware

Equipment for editing motion picture footage is quite specialized, and like any other specialized tool, it can be expensive. Unless your organization is already equipped for this sort of work, you should consider one of two alternate approaches. The first would be to rent the equipment yourself from a motion picture equipment supplier. This is practical, however, only if you or someone in your organization has some experience in editing film.

If not, your best approach would be to hire a film editor. If you hired an outside cinematographer, the chances are pretty good that he or she can give you the name of an editor who is equipped for the job.

Editor works at a flatbed editing table.

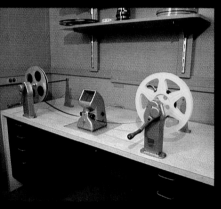

Rewinds and viewer

Sound reader

Tape splicer

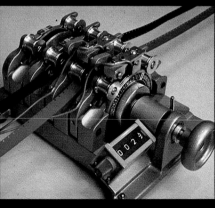

Sync block

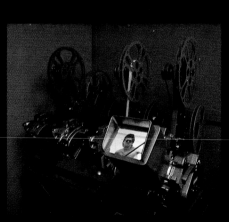

Moviola

Trim bin

165

If your shooting was done from a script and a comprehensive shot list, the process of editing is more or less straightforward. If you don't plan to edit the footage yourself, an experienced editor can accomplish the job easily if he or she is given a script, a shot list, and the camera log.

If, on the other hand, the shooting included a lot of opportunity shots or some unscripted interviews, the editing job will be a little more involved. Editing an unscripted interview is time consuming because your subject probably was allowed to talk at length. Your chore is to shorten the interview to a practical length. The easiest way to accomplish this is to transcribe the original ¼-inch sound track; then you can "edit" on paper. The result is the equivalent of a script against which the film can be edited. Naturally, you must check your "paper edit" against the film footage to see if the sound track can be cut to conform to your wishes.

All the steps touched on so far are common to both normal motion picture editing and editing for a multi-image presentation. However, since the movie film you're editing is to become part of an overall show, you'll have to provide your editor with a few additional facts. This is especially true if the person doing the work has never edited motion picture film for this type of show.

To begin with, most professional editors perform more than a purely mechanical service. They don't simply "cut and paste," even when working from a tightly written script. They're accustomed to making — or at least recommending — aesthetic decisions that can be of value in holding a scene together or making one scene work with the next. As a result, an editor may try to make a film segment tell the whole story by itself rather than integrate it with slide images that are on the screen at the same time. So it's imperative that the editor know just how the film segment is to work as a part of the whole.

Of equal importance will be a knowledge of **where** on the overall screen the film segment will be projected. If, for example, the overall screen area measures 10 x 30 feet (3 x 9 m), and the film segment is going to fill an area within the left-hand third of the screen, the editor should be aware of that fact. Then he or she can orient the film image toward the center of the overall screen area. (Of course, this is a fact that should have been considered during the shooting of the film, so the basic shooting angle should be correct. However, if there is to be more than one film image and you plan to use two or three motion picture projectors aimed at different areas on the overall screen, you'll obviously have a variety of shot angles.)

The editor should also be told how much black leader is required between the film segments, as well as which segments will be used on which projector when you plan to use more than one motion picture projector. Just as an editor or conformer produces "A" and "B" rolls in a checkerboard fashion for printing purposes, your editor will produce two or more "A" and "B" roll sets in a checkerboard fashion when two or more projectors are going to be used.

Another fact your editor should consider is the type of sound track you'll want on the release prints — either optical or magnetic. Although this choice doesn't affect his or her work directly, that information must be included in the instructions to the lab that handles the final production of your film segments.

At this point in the process, your motion picture sequences should be edited, and the accompanying sound track (if there is to be one) should be mixed to incorporate the sync or voice-over narration, the sound effects, and the music. You're almost ready to get the original camera negative conformed to the work print (by the editor or lab) so it can be sent for any necessary optical effects and a check print or answer print. But first there's an additional step you must take. As we mentioned when discussing the organization of the total production, the longest path on your critical-path chart will most likely include the completion of the motion picture sequences. To get release prints in time to meet your show deadline, the content of the film segments will probably be locked in long before most of your other visual material is available. As a result, you'll want to be sure that the film sequences are right before they're conformed and sent off to the lab for final release prints. Now is the time to check

them against your storyboard (if you've used one), and to get final approval from your client. This is a critical factor, since — very often — clients looking at a work print or interlock don't always realize that they are being asked for a final approval so early in the production schedule.

Once the film segments have been approved and you've had the negative conformed, the edited film is sent to a lab. Now the production process leaves your immediate control for a brief period. But "out of sight" can't become "out of mind." When you developed your critical-path chart, you considered the time your lab needed to create optical effects, an optical track, a check print, an answer print, color correction, another answer print, and finally, a release print. Now you must follow up on that schedule. While you're involved in the other aspects of your show, you must also keep tabs on each step as your film segments proceed toward the final release print. If you don't, you could reach your show deadline without film sequences. (It's not uncommon — though certainly undesirable — to find yourself using an unacceptable answer print for rehearsals or even client previews.) If you contacted a lab and received realistic estimates of the time necessary for each phase of the operation, **and if you follow up with the lab as each checkpoint arrives,** you will have a fairly reasonable assurance that your film segments will be done correctly and on time.

167

Editing And Preparing Slides

It may seem as if we've been ignoring the principal medium in a multi-image presentation—the slides. We've considered screens, seating, script, storyboards, sound, and motion picture sequences. But we haven't said much about the medium that makes up 80 percent or more of your presentation.

Now you're going to begin selecting and preparing those all-important slides. So now you should begin to feel as if things are finally happening. If your show isn't too complex or if it has been tightly storyboarded, this task will be relatively simple. In fact, it may require little more than selecting slides with the right exposures and dropping them in their trays.

On the other hand, you face a demanding assignment if your show required a lot of candid photography, shot from a very rough storyboard or from script call-outs; and if some of your slides were shot for very specific progressive-action sequences; and if you've shot art slides for animation or progressive-disclosure sequences.

If any of these considerations apply, you may have to choose from several thousand slides covering a wide range of subjects. So your first task is to impose some kind of organization on this mountain of slides. The question is, where—and how—do you begin?

Experienced producers usually experiment with several different procedures until they find one that works best for them. What we're about to describe, therefore, shouldn't be interpreted as the only way to approach this selection process. It is a basic approach that you can modify to suit your needs.

168

First of all, we recommend you have all your slides returned from processing mounted in cardboard or light, plastic mounts—even if you plan to remount the presentation slides later. This is a matter of convenience. You may save some money having slides returned from processing uncut, but you lose that money quickly as you sort and edit slides from strips of film.

Next, you must separate your slides into categories. If you or your photographer logged the film as it was shot (and the roll numbers remained with the film through processing and mounting), you'll be able to break the slides down into rough categories even before you open the boxes. For example, if rolls 1 through 30 were shot at an industrial facility in Texas and rolls 31 through 40 were shot in metropolitan Chicago, you've already imposed some organization on your slides. If, on the other hand, no logs were kept, you'll be starting at the beginning.

In organizing your slides for a show, you'll want to set two—possibly three—goals right at the outset.

First, of course, you'll want to separate the slides according to subject. You can create broad subject categories or more specific categories. We usually avoid getting too specific during the first culling. Your principal objective during this initial sorting should be to make the slides readily accessible by category. Categories that are too restrictive can leave you with a problem almost as big as the one you set out to solve. (Keep in mind that we're talking about **categorizing** slides for use in a current production; we're not talking about **cataloging** slides to be filed away for future use.)

Second, you'll want to evaluate exposure and pull out the best of each shot.

Third, you'll want to gather any slides that are to be cropped or masked for special effects. For example, if your show will incorporate vertical images and other special shapes, you'll want to separate the slides shot for that purpose.

So much, then, for **what** you want to accomplish at this point. Now for the **how. View your slides.** That seems obvious, but unfortunately, it's a step many people disregard completely or, at the very least, don't perform correctly. By correctly we mean under the right conditions—in a room as dark as the one in which the show will be viewed, with a projected image size as large as possible.

A stack loader simplifies the process of editing large numbers of slides.

Plastic sleeves provide an effective means of temporarily storing slides to be used in a presentation. During production, slides can be arranged in sequence on a large illuminator.

We stress duplicating actual presentation conditions for an important reason: The amount of light that reaches a screen is determined by many things such as projection distance, image size, image density, and the light output of the projector lens. In other words, a slide that looks perfectly exposed in a totally dark room on a small screen may seem badly underexposed when it's projected on a large screen with some ambient light in the room. Conversely, a slide that looks great on a sorting table may look badly overexposed when viewed full size on a screen in a darkened room. So don't depend on a back-lit sorting table or illuminator. They're excellent for sorting and laying out slides in a sequence, but not very helpful when it comes to evaluating exposure, color balance, composition, focus, and the facial expressions on your subjects. In addition, illuminators won't help you discover such potential problems as processing stains or camera scratches—two gremlins that are seldom visible until a slide is projected on a large screen.

We've found that a stack loader such as the Kodak Carousel stack loader is the most effective means of viewing large numbers of slides in the shortest period of time. This device holds up to 36 mounted slides at a time and enables you to quickly remove bad—or unwanted—slides while viewing them. Some producers prefer to load all their slides in trays and view them using two or three projectors. While this approach does offer one or two advantages (the slides can be kept in trays identified by category and the trays can be reversed to review previous slides), it also requires more trays and more preparation.

Note instructions for special masking or mounting directly on slide mount.

Once all slides have been viewed, evaluated, and separated into categories, they can be kept temporarily in labeled boxes. The best exposures for each shot can be inserted in labeled 8½ x 11-inch (216 x 279 mm) slide file sheets for quick review and easy access.

When you're producing a multi-image show that uses full-aperture images, the selection of slides is fairly straightforward. This is the time when a large sorting table or illuminator is useful. Sequences of slides can be arranged in rows or columns, depending upon the number of screens or screen areas in the presentation.

If your presentation requires selective masking or sandwiching of slides, you should make notes on the slide mounts to indicate any special treatment required. The effects you can create by using special mounts and masks or by using animation stand techniques are almost limitless. Although we'll attempt to touch upon some of them, it would be virtually impossible to cover them all within the confines of this book.

Special Mounts

The simplest method of creating different visual shapes on the screen is to use commercially available slide mounts. These are readily available in a number of basic shapes — squares, circles, triangles, and a variety of rectangles. Other shapes can be specially ordered if sufficient time is available.

Masks

You can use two basic types of slide masks. The first is a paper mask that is mounted with a transparency and then joined between two pieces of glass in a metal slide frame. A number of manufacturers produce a variety of paper masks. Unfortunately, many of these masks have two major drawbacks. First, because they are die-cut from either plain paper or foil-covered paper, they tend to have rough edges that show up on the screen. Second, and more important, paper masks do not provide a means for automatically registering one image on top of another.

A more versatile and sophisticated method of masking involves producing custom masks on Kodalith film. They're more versatile than paper masks because they can be created in any shape or size desired. And they're more sophisticated because they mount directly on the sprocket-hole registration pins in a standard slide mount, such as those produced by Wess Plastic, Inc., of Farmingdale, New York.

(above) Plastic mounts are available in a variety of configurations. (below) Special masks can be purchased or created using Kodalith film.

171

Effective graphic and type slides can be produced from easy-to-make art. Many producers take advantage of the convenience of transfer type.

The Magic Of **Kodalith** Film

Kodalith ortho film 6556, type 3, is a high-contrast black-and-white film that is available in 100-foot rolls from graphic arts dealers. Its "magic" is the ease with which it can be shot and processed and the wide range of uses it has in multi-image production. Kodalith film is especially useful for creating custom masks—a process we'll describe shortly. It also can be used to produce crisp graphic or copy slides that can be used as they are—in black and white—or with the addition of color by sandwiching them with gels or dyeing them with commercially available water color dyes. And, as we mentioned earlier, copy or graphic slides produced on Kodalith film can be used to produce burn-ins on other slides.

To produce a copy or graphic slide on high-contrast film, you begin with a piece of art that's prepared in black and white. There are, however, two facts to keep in mind when preparing the art. First, Kodalith film is a negative film. Any copy or graphic that you want to appear as white (which can then be colored with dyes or gels) must be black on the artwork. Second, Kodalith ortho film 6556 is an orthochromatic film. As such, it is sensitive to many shades of blue and "blind" to all shades of red. In other words, certain shades of blue will show up as white on a Kodalith film reproduction, while reds will reproduce as black.

In preparing the art for a copy slide, you can use type set by a commercial typesetter or you can do it yourself with any of the "rub-down" (transfer) lettering products listed in the appendix. Keep in mind that because Kodalith film is selective in what it "sees," it's extremely useful when you want to produce progressive-disclosure slides—a series of slides on which visual information is added in steps.

A plain piece of white paper can be used to cover the words or graphics for subsequent steps while you're shooting the first—or base—step. Then, as each subsequent step is shot, the paper can be moved to reveal additional information. If the paper is kept relatively flat against the artwork and lighted to avoid a shadow along its edge, it will not show up in the finished slide.

Although we mentioned this earlier, it's worth repeating that art prepared for copying should be produced on pin-registration boards to assure squareness, proper alignment and, especially, accurate registration for animation or progressive-disclosure sequences.

To produce a special mask on Kodalith ortho film, you begin in much the same way as that described for producing a copy or graphics slide. In this case, however, your black-and-white artwork is nothing more than a black shape on a white background. (You could also use a red "block-out" adhesive film instead of black on your artwork. The advantage these films offer is that they are transparent and can be easily cut to produce clean, sharp edges.) Once your artwork has been photographed, and the orthochromatic Kodalith film has been processed, you'll have a series of negatives that contain perfectly clear—and precisely repeatable—"windows" or masks. Because these masks drop onto the registration pins in plastic slide mounts, you'll be able to maintain accurate registration on the screen.

One of the distinct advantages this technique offers is the ability to project several images on a single screen area. If, for example, four projectors are aimed at the same screen area, the slide for each projector can be sandwiched with an appropriately positioned Kodalith film mask, creating a different image in each quarter of the screen. (See illustration.)

Blocking film, available from art supply stores, also can be used to create special masks using Kodalith film.

Special screen configurations, such as the one illustrated below, can be created using masks produced with Kodalith film.

Masks for multi-projector panoramas also can be produced with Kodalith film. However, because this type of "seamless" mask is meant to create a graduated density on one or more edges of the frame, both the preparation of the artwork and the processing of the film vary from what we described above. The best way to produce the evenly graduated artwork needed to produce this kind of mask is with an airbrush, black ink, and a very steady hand. Then, after the art has been shot on Kodalith film, it must be processed with continuous-tone developer (such as Kodak Dektol, D-76, or equivalent) instead of Kodalith developer. The illustrations at the left show a set of masks for a multi-projector panorama and the image they produce on the screen when they've been sandwiched in slide mounts. Keep in mind that the slides for such a panorama must be shot with precision for them to register in this manner.

(Because the processing of Kodalith film for seamless masks must be precise, some producers prefer to use either a fine-grain positive film or Ektachrome film.)

Frames for motion picture images also can be created using Kodalith film. While the use of motion pictures in multi-image presentations has always been effective, its major aesthetic shortcoming was that if the image did not fill the screen, its soft edges lacked the finished look of slide images. A Kodalith film frame eliminates that drawback. Produced the same way as a Kodalith film mask, this frame can be sized to burn out the soft edge of the film image. Although the accompanying illustration shows a simple solid-line frame, you could create one as ornate or complex as your imagination can make it. When you're preparing the art for a motion picture frame, keep in mind that its size will be influenced by the relative focal lengths of the slide projector lens and the motion picture projector lens.

Other special effects can also be created using Kodalith film. Earlier we mentioned that magazines and brochures are visually appealing because their pages are designed as a unit. Well, nearly anything an art director can do with pictures, shapes, and type on paper can also be done on the screen with images and Kodalith film masks or frames. For instance, would you like your framed motion picture image to "float" within a larger slide image? You can do it by sandwiching your slide image with a Kodalith film mask that has a black area corresponding to the framed motion picture area. You can put one slide image within another in the same manner. You also can sandwich slides with Kodalith film masks in the shape of typographical characters and spell out short words across the screen in pictures. The possibilities are endless.

Using An Animation Stand

With the development of faster and more sophisticated programming equipment, the achievement of motion-picture-like animation with slides became a reality. However, the fact that slides can be flashed across the screen with such speed is of little value if the images are not kept in precise registration. This is where a good animation stand plays a key role. Because you have the capability to move artwork in any direction—precisely—you can shoot slides of the image at any point within the frame, and at any focus, to create the specific visual effect you're after.

This precise registration capability, coupled with the transmitted-light feature available on most good animation stands, enables you to duplicate an existing slide image in a reduced size so it can be positioned accurately anywhere in the frame area. If you intend to duplicate slides that fill only a portion of the frame area, however, we suggest you sandwich the resulting slide with a Kodalith film mask. Otherwise, the density of the dark area on the duplicate slide will not be sufficient to block light from passing through the dark portion of the frame.

The creative use of masks opens up a wide range of possibilities to a producer.

175

Richard Shipps programs an animation sequence. (Photo: DD&B Studios)

Richard Shipps: What You See Is What You Think You See

When Richard Shipps plans animation sequences for a multi-image presentation, he's thinking as much about what people won't see as he is about what they will.

"The most fascinating aspect about producing animated sequences," says Shipps, founder and president of DD&B Studios in Detroit, "is that the audience does part of the work. As a producer, you have to design a sequence so the audience can fill in the gaps between the images on the screen. Animation is not so much what people see as what they think they see."

In short, the total effect exceeds the sum of the parts.

In a word, illusion.

Shipps and DD&B have been creating extremely effective and widely acclaimed multi-image illusions since the studio's founding in 1974. Among the clients for his highly sophisticated productions have been the Oldsmobile and GMC Truck Divisions of General Motors, IBM, and a company with special reason for delighting in Shipp's visual magic—Audio Visual Laboratories, Inc., a multi-image programmer manufacturer. For example, as part of a presentation for AVL, produced to promote its Eagle programmer, Shipps created an animated sequence in which an eagle seems to fly across the screen, from one edge to the other, then loop back into center screen. But as Shipps would be quick to point out, the eagle doesn't fly. The viewers create the flying in their minds. The illusion just takes advantage of a natural human tendency.

"People are very quick to accept a repetitive action," he says. "Anything that happens repeatedly and continually will be taken as fact, and people will begin to form judgments based on these repeated observations. When they see a sequence of images, repeated over and over with only the slightest variation, they forget they're looking at individual images and form the conclusion that they're looking at motion."

Shipps sees multi-image production as a way to extend the boundaries of the art. "With multi-image techniques and equipment," he says, "you have more control over the process, because you can constantly adjust and change this time reference. You can run sequential images at a rate of 10 or 20 or 30 images a second, depending on your equipment capability. And you can vary the interval between the images as well, creating instantaneous changes or two-, four-, 16- or 32-second changes. As a producer, your options are almost limitless, and that's what makes the field so interesting."

Creating these animated sequences is a process that blends planning with spontaneity. Shipps begins his planning by breaking down a total show into its individual screen effects. Then he blocks out the time for each effect. A generalized storyboard illustrates the action that is to take place in each sequence, but Shipps avoids detailed storyboarding of individual steps. "The creation of the individual steps should occur spontaneously," he says, "when you're working at the programmer. This ability to be creative spontaneously, of course, is an advantage the motion picture animator doesn't have."

The general storyboards go to DD&B's five-board art department, where the camera-ready artwork is prepared. AVL's "flying eagle" was created using traditional sequential tissues, he says. A Xerox copier was used to create acetates, which then were painted. The animation cels were then aerial imaged into other kinds of backgrounds.

All camera work is also performed at DD&B, using a Forox camera. Slides are processed in-house too, using two color processors. "We're geared to spontaneous production," Shipps says of his equipment and full-time staff of 15.

Shipps says he creates more art and shoots more slides than he'll eventually use because it's the additional materials that allow him to be inventive. "I work at the programmer a lot, adjusting timing and art. If something doesn't work, we pull it out and insert something else. We'll continue to create material until a sequence works, both as an integral unit and as a piece of the whole." By working in this way, he adds, "I can create effects far more precisely than I could if I tried to plan for them."

"Most good multi-image presentations come together spontaneously," he says. "The final slides in the gate are, in many cases, the ones you've inserted during the last 24 hours of production. That's one of the attractive aspects of working in multi-image. You can create, learn from what you create, then use those lessons to improve the original creation."

To illustrate his point, Shipps offers an example: "Let's say you've created a wipe or a sweep using 12 projectors aimed at different screen areas, and you've plotted the individual positions so you have a smooth and fluid motion. You run the sequence and you like the effect but you don't like its length; it's too long, there's too much time between images. With multi-image you can go back and redesign art so you can put two images side-by-side on each of the 12 frames. What you've done in effect is to create a 24-step move. Even though it's seen as coming out of 12 images, if the images are presented sequentially, you can fool the eyes of the viewers into thinking they're seeing 24 individual increments."

It's this ability to deceive the viewer's eye—to create illusions—that Shipps finds to be the most intriguing part of multi-image production. "People watching multi-image animation know they're looking at still images. They know it's just a series of single slides they see on the screen, yet in their minds they perceive motion. Their reaction is, 'That can't be done.' "

It's this apparent contradiction between what the viewer knows and what the viewer sees, Shipps believes, that creates intense viewer involvement with multi-image animation. This "theory of viewer involvement" is firmly grounded in the psychological observation that people are more attentive to unfinished tasks.

"People become absorbed in multi-image presentations because they subconsciously try to see images as stills. They try to get as much information from each image as they can, but before they know it the image has been replaced by another, so they try harder the next time to perceive the information on the new slide. The process continues with each slide change, with the viewers trying harder and harder to extract information. And the harder they try, the greater their attention and retention."

Of course, this observation is especially true when viewers are trying to extract information from images created by Shipps, a master of multi-image illusion.

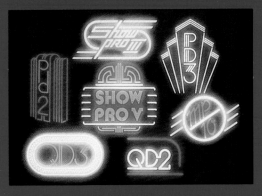

Images from three DD&B productions: (top) From **Five Finger Exercises,** produced for Audio Visual Laboratories, Inc. (center) From **Free Fall,** produced for Audio Visual Laboratories, Inc. (bottom) From **Oldsmobile Omega Show,** produced for Oldsmobile Division, General Motors Corp. (Photos: DD&B Studios)

177

PUTTING IT
ALL TOGETHER

Selecting The
Basic Equipment

By now you've probably realized that selecting equipment for a multi-image presentation is very much like buying a new car. First you have to look at what you need — compact, economy-price equipment or full-size, full-feature luxury equipment, or something in between. Then you have to look at the market-place — where you can choose from among many models, features, and price ranges.

Given all the variables, choosing the equipment you need is difficult enough. It becomes doubly difficult if you start searching for the right combination of equipment. Because the truth is, there's no "right" combination. Just as there's no one car that's perfect for you, there's no one combination of multi-image equipment that's perfect for all of your needs.

So forget about perfect combinations and think instead of optimum combinations — those pieces of equipment that will serve you best most of the time.

In this chapter we'll look at a variety of products that represent different levels of sophistication in audiovisual and multi-image production. Obviously, it isn't our intention to duplicate the contents of the most comprehensive equipment catalog — the **NAVA Audio-Visual Equipment Directory** (published by the National Audio-Visual Association). Nor is it our intention to recommend any one piece of equipment as being the most sophisticated, or the simplest to operate, or the most reliable. We will, however, attempt to highlight those products that — in our judgment — brought innovation and increased flexibility to the industry. At the same time, we'll attempt to show some of the ancillary pieces of equipment that have evolved as the industry has matured.

Slide Projectors

The slide projector has been called "that ubiquitous device present in every class-room, training room, and board room." You can add to that definition that it's also the basic piece of equipment for multi-image production. Many different makes and models of slide projectors are available, but the Kodak Ektagraphic slide projector has become very popular in the audiovisual industry. Kodak markets a number of models that range from a basic projector through units that incorporate a built-in self-timer and an automatic focus-ing device (with an autofocus on-off switch) to units that feature random-access capability and a dark-screen shutter latch that eliminates the need for opaque slides. Other manufacturers market projectors that use the same basic slide-advance mechanism coupled with a variety of high-intensity light sources.

Two other points about slide projectors are worth noting:

• We recommend that you use the same type of projection lamp for both programming and presentation, regardless of whether it's a low-voltage, line-voltage, or arc lamp. Keep in mind that low-voltage lamps won't turn on and off as rapidly as line-voltage lamps, because they're made with heavier filament wire. This can cause problems, particularly if you're projecting fast-paced animation sequences. In addition, if you use line-voltage lamps for the actual presentation and low-voltage lamps during programming, you'll notice that the smoothness or timing of your animation sequences may be off.

• Slide trays holding up to 140 slides are available, but we recommend you use 80-slide Kodak Ektagraphic universal slide trays, even if this requires tray changes during a presentation. The narrow-slotted, 140-slide trays are not made for use with glass- or plastic-mounted slides. Even if you use cardboard mounts, the high number of frequent slide changes greatly increases the risk of slide hang-ups.

Projector Racks

Not too long after people started using multiple projectors, it became obvious that there had to be a method of safely stacking two or more projectors to conserve space and reduce keystoning. Initially, producers confronted with this problem simply built shelves to fill the need. Since then, a number of manufacturers have developed a variety of specialized projector racks. Some racks are simple in design and relatively inexpensive. Others are more complex and enable the user to make fine adjustments for critical alignment of the projectors. More recent models are incorporated into sturdy shipping cases. With these, two or three projectors can be shipped with their lenses and slide trays in place.

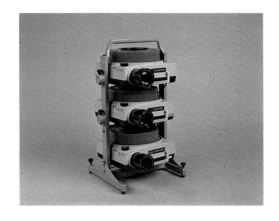

Two- and three-projector racks are available in designs similar to the one shown here, manufactured by Columbia Scientific Industries.

As part of its Director-24 System, Spindler & Sauppé markets four-projector racks that incorporate a dissolve module.

Motion Picture Projectors

While most audiovisual and multi-image producers and users probably consider only one type of slide projector, these same people use many different makes and models of motion picture projectors. Not counting super 8 and 35 mm projectors, which have relatively limited application in the multi-image field, we're left with no less than 15 manufacturers of 16 mm projection equipment. And when you consider the number of models produced by each manufacturer, you find your number of possible choices comes to almost 30. These units offer many features ranging from self-threading capability, to continuous-loop operation, to combination optical/magnetic sound.

From the point of view of multi-image production, however, few of these features are of great value. The important features for a multi-image producer are a projector's capability to start and stop consistently without losing a loop and its capability to come up to speed quickly and smoothly. Beyond these, any other features a multi-image producer would want on a 16 mm projector would have to be added by the producer. These custom options might include miniature switches or sensors that stop the projector when a notch on the edge of the film or a foil tab goes past them, or a synchronous or stepping motor that accurately controls the speed.

Projection Lenses

The lenses you'll need for your slide projectors and motion picture projectors will depend on both the projection distance and the image size you'll want at the presentation site. However, if you are going to be making presentations under a variety of conditions, you may find it necessary to have a range of sizes. This could become expensive, depending upon how many slide projectors you'll be using. On the other hand, renting 15 or 20 lenses in several cities across the country might prove equally expensive — if it's even possible.

So we recommend that for your slide projectors you have a set of 3-inch (76 mm) lenses (primarily for rear projection), a set of 4- to 6-inch (102 to 152 mm) zoom lenses (for medium projection distances) and a set of 8½-inch or 9-inch (229 mm) lenses (for long projection distances). For 16 mm projectors, you should consider a 1½-inch (38 mm) lens, a 2-inch lens with a bifocal converter, and a 3-inch or 4-inch lens.

Should you need them for special applications, lenses are made in standard sizes from 1-inch to 19-inch (483 mm) focal lengths. There are also special lenses designed for short-throw rear projection that incorporate front-surface mirrors. These lenses offer two benefits: First, they enable you to position the lenses of two projectors right next to each other (to reduce keystoning); and second, they automatically reverse the image, making it unnecessary to change the orientation of the slides in the trays.

Two relatively new entries in the marketplace are:

• a series of superwide-angle lenses for slide projectors; these lenses provide large images over a short projection distance without distortion;

• a motorized zoom lens that permits changing the size of the projected image without a focus shift.

The latter lens does have one weakness worth noting. Since it is basically a modified camera lens, it requires a higher level — "hotter" — light source to put the same amount of illumination on the screen compared with a normal projection lens.

Reversing Film Images

If you're using 16 mm sound film in a rear-projection presentation, you may also need a mirror lens for your motion picture projector. Unlike slides, which can be reversed in a projector for direct rear projection, 16 mm sound film must be projected in the conventional manner for the sound track to be heard.

There are three ways of reversing the sound film image:

You can print your sequences on double-perforated silent film (which can be threaded in a projector in a reversed position) and transfer your sound to your audiotape. Maintaining proper sound/image synchronization is difficult with this procedure, however, unless you have expensive stepping motors for your projectors.

You can also have a reversed print made of your film. This, however, can be extremely expensive.

The most practical solution is to use mirrors for folding the projected image or a mirror lens, which accomplishes the same purpose.

Sound Equipment

The sound system you put together for a multi-image presentation is apt to vary with the complexity of the presentation itself, the number of input sources, and the size of the room where the show will be presented. Frequently, producers will elect to bring their own tape deck and then rely upon a reputable audio house to provide amplifiers, mixing boards, and loudspeakers on a rental basis. This enables them to take advantage of equipment that is sized specifically for the presentation room. Other producers choose to ship a complete sound system with the rest of their show equipment. This gives them the benefit of working with a "known" quality.

Four-track tape units, such as the TEAC 40-4 shown here, are popular with multi-image producers because they offer stereo sound plus two additional tracks for synchronizing cues.

The basic element of the sound system for a multi-image presentation is the tape deck, and the number of makes and models available is staggering. They range from simple cassette units with or without built-in synchronization capabilities to elaborate solid-state eight-track units. In between are sophisticated cassette decks and continuous-loop cartridge units that rival even the best reel-to-reel units for reliability and quality. Although some of these new cassette decks are finding increased application among multi-image producers, most still rely on solid-state reel-to-reel equipment.

Not too long ago, producers were content to use stereo tape decks to play back the audio portion of their multi-image presentations. This had its limitations, however, because one track was needed to carry programming cues or digital information, so the presentation's sound track was confined to monaural sound. (Some producers modified their decks by adding a separate playback head for the cue track; this freed the two primary tracks for stereo sound.) Later, smaller four-track decks were developed for quadraphonic sound. While this equipment didn't meet with a great deal of acceptance for that application (the fad died out along with many of the products), it did become popular with multi-image producers. They saw it as a means of providing high-quality stereo sound with not one, but two additional tracks for cueing information.

Some producers, when putting together a complex sound track for a sophisticated presentation—especially if it's to be permanently installed—take advantage of eight-track decks. This equipment enables them to isolate—and play back—specific parts of their sound track from loudspeakers in different areas of the presentation site.

If your presentation is going to include motion picture sequences, and if you plan to use the sound track from the film footage, you'll need a small mixing board to control the sound going into your amplifier from both the tape deck and the movie projector. Although many of the commercially available audio mixers are extremely sophisticated — primarily because they're designed for mixing down complex multi-track sound tracks — your requirements for presentation sound are basic. So don't give in to the temptation to buy more than you need for the application.

The balance of the equipment you'll need for your sound system — should you decide that you'd rather buy than rent — will include a commercial-quality power amplifier, loudspeakers, microphones, a multi-band acoustic equalizer (to compensate for poor acoustics at the presentation site), and possibly an audio time-delay unit — with an extra power amplifier and a pair of loudspeakers for the rear channels — to add spaciousness to the sound.

Choosing an amplifier and loudspeakers is an extremely subjective undertaking. What sounds acceptable to one person might sound too bassy for another, or too tinny for yet another. Sound comes down to a matter of personal taste and experience. The only recommendation we can safely make about the selection of audio equipment is to choose a power amplifier with sufficient power and choose loudspeakers with sufficient frequency response, power-handling capacity, and horizontal dispersion to "fill" the largest room in which you're likely to be putting on your multi-image presentation. It's much better to have a sound system with more power than you need than to have one that isn't capable of putting out low-distortion sound at a sufficient volume level.

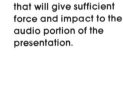

Choose a power amplifier that will give sufficient force and impact to the audio portion of the presentation.

Compact audio mixers enable a producer to control the sound levels from microphones, tape decks and motion picture projectors.

(top) One of the first commercially available dissolve units was introduced by Eastman Kodak Company in the early 1960s. (bottom) In 1969, Spindler & Sauppé introduced the first multi-speed dissolve control.

Dissolve Controls

The ability to blend one image into another stands as a major development in efforts to give audiovisual presentations a look of professionalism. And for a few short years after the Kodak Carousel dissolve controls were introduced in 1964, they were lauded as the greatest thing that had happened to the industry since the advent of the remote control projector. As the 1960s were about to close, Spindler & Sauppé, Inc., introduced a dissolve control that was capable of not one, but five functions. Then, as the 1970s began, other manufacturers introduced dissolve controls capable of performing even more elaborate functions such as laps, cuts, and flashing.

It was the increased capabilities of dissolve controls—as much as any other single factor—that prompted the development of sophisticated programming equipment. While a simple dissolve control could be synchronized with a sound track in the same way a single projector was synchronized, a multi-function dissolve unit required something more than a single-function tone to operate it for anything other than a manually controlled presentation. So, while there are a large number of dissolve controls on the market—some of which are even capable of generating multiple-frequency or variable-duration tones for synchronization purposes—the more sophisticated dissolve units have their control functions incorporated into programmers.

The Kodak Ektagraphic EC-K solid state dissolve control.

(far left) Mark IV multi-function dissolve system control module, manufactured by Audio Visual Laboratories, Inc. (left) Audio Visual Laboratories' Mark VII, the first dissolve control module to link three projectors.

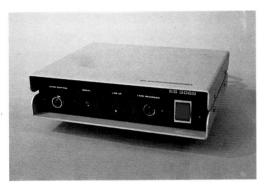

(far left) Showslide 3069 dissolve control, manufactured by Electrosonic Systems Inc. (left) The Entré 7600A variable-rate dissolve-programmer.

(far left) The QD 3 computerized three-projector playback dissolve module, manufactured by Audio Visual Laboratories, Inc. (left) Spindler & Sauppé's ADX Decoder controls up to four D-24 dissolve modules.

The Wollensak/3M AV-44 multi-function dissolve unit can be controlled manually or through a memory programmer.

Note: For more information on the products shown on this page, see your local audiovisual dealer or the respective manufacturers.

Programming

The Difference Equipment Makes (And Doesn't Make)

In the introduction to this book we stressed the importance of modern programming equipment in the production of multi-image presentations. We pointed out that the programmer is the basic multi-image production tool, and that the sophistication of a programmer wields the greatest influence over what a producer can and cannot do on the screen.

Now let's examine those statements more closely, to discover the difference a programmer makes—and doesn't make—in programming a presentation.

First, by analogy, look at what it means to regard a programmer as a basic tool. A programmer is to a multi-image producer what a saw is to a carpenter—a tool used "to help make separate pieces fit together." Without their respective tools, neither the carpenter nor the producer could construct their products according to a plan.

In the carpenter's case, he or she would have to use lumber in whatever sizes were available, trying to find enough pieces of equal length and width to construct walls, floors, and ceilings. Or the carpenter could try to break or chop the lumber to specific dimensions. But either way, the task would be difficult and the resulting structure would be ragged.

In much the same way a producer could produce a multi-image presentation without using a programmer. The producer could set the multiple projectors on automatic advance and hope the slide changes coincided with the narration or music; or the producer could try to control the projectors using a makeshift keyboard constructed of remote-control units. Neither approach is impossible, but like the carpenter, the producer would find the labors difficult and the presentation haphazard, with plenty of rough edges.

So a programmer, like a saw, is a tool of precision, of craftsmanship. It enables a producer to create according to a plan, to work with predictable results—**to exercise control over materials.**

But a programmer is not the plan and it is not the result. Those are the products of the craftsman—the producer. It's important to keep this distinction in mind, because later in this chapter you'll see that a programmer is only as effective as its operator.

What's In The Future?

Predicting the future is a risky business, especially when the predictions are (a) to be contained in a book and (b) focused on an area where the momentum of technological change is accelerating.

So we'll hedge our bets. We won't speculate about specific product features you're likely to see. Instead, we'll talk about trends we think will develop in the industry.

First of all, we think that while you'll see continued refinement of top-of-the-line programmers, you won't see any radical equipment breakthroughs. As Audio Visual Laboratories has demonstrated with its Eagle, multi-image programmers have reached the same level of sophistication as the minicomputers used for payroll accounting and inventory control. If programming equipment itself is to become more sophisticated, it will have to wait for breakthroughs in the minicomputer industry.

But there's an even more practical reason for believing that equipment manufacturers will shift their emphasis from greater capacity to greater refinement. The market for programmers capable of controlling dozens of projectors with tens of thousands of cues is limited. Because of this, we believe the

future of the ultrasophisticated multi-image programmer will most likely repeat the history of the full-size computer. There will be more emphasis placed on simplifying programmer operation and greater effort put into the development of programming software.

The history of the computer industry also provides clues to two other trends we should see in programmer development and marketing: first a forceful move to miniaturization of equipment, followed by the spread of what can be called "grassroots programming." Together these trends will promote the democratization of technology — programmers for everyone, from the small producer to the large.

This movement will mean more products, of greater sophistication, for producers using between two and 10 projectors in their presentations. Some of these products may be multifunctional — for example, units that combine programming and dissolve functions. Other products may reflect greater specialization and simplification — for example, program "readers" will be used at presentations, freeing the more sophisticated programmer for its proper function of program creation.

There's another distinction that should be made when looking at the sophistication of programming equipment. Just as there are hand-held saws, portable power saws, and heavy-duty, multi-purpose bench saws, so there are basic programmers, advanced-function programmers, and ultrasophisticated programmers. All perform the same basic functions; however, some — whether saw or programmer — give an operator greater flexibility and more convenient operation. For instance, the more advanced a programmer becomes, the greater its capability to vary and com-

bine equipment functions, the greater the number of commands it can handle, and the easier it is for an operator to create special effects.

So technology makes a difference — to a point. **It makes programming more convenient and more accessible to more people, and it gives an operator greater latitude in which to experiment and innovate.** But technology isn't a substitute for creativity or talent. That, once again, comes from the craftsman.

The Growth Of Programmer Capability

One of the inherent challenges in audiovisual production has been to synchronize the audio and the visual — to interlock sight and sound into a single impression. At first producers achieved this linking by using their eyes, ears, and hands — they saw or heard a cue and manually triggered a slide change. Of course, this method was never truly satisfactory, so the producers sought more sophisticated, more responsive, and more reliable means with which to synchronize slide changes with a sound track.

The first equipment developed for this purpose used a single-frequency tone that advanced one projector or a pair of projectors connected to a dissolve control. For producers, it was better than

The simplified panel of the DuKane Custom Electronic Programmer enabled producers to control up to eight projectors or projector banks.

manual synchronization, but they readily recognized its limitations. They wanted to work with more than a single projector or a dissolve-linked pair. They wanted to create shows using many projectors — perhaps as many as eight.

Again, manufacturers offered solutions. One approach — the earlier and simpler — was to use a number of different frequency tones, one for each projector or pair of projectors. On paper, the approach was feasible, but in application it just didn't work reliably. The tape recorders of the day were unable to maintain constant speeds, so the frequency of the tones would shift, causing false tripping and other equipment malfunctions.

A second approach relied on a standard punched paper-tape reader to control projectors or pairs of projectors. This system proved more reliable and became the heart of the DuKane Corporation's Custom Electronic Programmer.

Paper-tape programming was everything producers were looking for — at least for awhile. The programmer provided eight channels — or paths along which commands traveled from programmer to projection equipment. This allowed a producer to send single commands to eight separate projectors or dissolve banks. It was with this equipment that multi-image production came of age. But it still had considerable growing to do.

That growth was stimulated by the introduction of multi-function dissolve units. When these products came on the market, producers realized they could use more than half of their eight-channel programming capability just to operate one unit. So for producers stretching the boundaries of multi-image production, the simple eight-channel programmer was no longer enough. Their need for greater capability and capacity gave birth to the current generation of programming equipment.

What's Best For You?

Experience tells us that by now 90 percent of the readers of this book are probably thinking about one question:

What's the best programmer on the market?

Our answer is sure to disappoint you: There is no best one.

Just as there's no best still camera, no best car, no best watch, no best golf ball, there's no best programmer **per se.** Of course, equipment manufacturers are ready to argue this point, but their claims and demonstrations don't change a fundamental fact of purchasing: **The best programmer on the market is the one that fits your needs best, both today and in the foreseeable future.**

The only way you can determine your present and future needs is to look at the requirements generated by current productions, then take an educated guess at how these requirements may change in the next one to three years. When examining these requirements, try to determine the maximum number of projectors you'll use, the probable functions you'll want in dissolve controls, and the length and complexity of your productions. That will give you an idea of the cue capacity you'll need.

After you've developed this information, look at the equipment on the market. You'll find that you'll have to choose from three basic types of programmers:

The simplest — and least expensive — are **tone-control programmers.** These are the more sophisticated descendants of the early single-tone programmers, now more reliable because of improvements in tape decks. Depending on the make and model tone-control programmer you select, you can produce six to 10 different tones. These tones are recorded on audiotape opposite your sound track, giving you the ability to activate six to 10 different projectors or dissolve functions. Your pre-

sentation requirements will determine how you use these tones. Using a 10-tone programmer as an example, you could send individual tones to 10 projectors, signaling them to advance their trays. Or you could send two tones to each of five dissolve/projector banks. One of these tones could activate the dissolve function, the other might create a fast cut. And, of course, you could divide the 10 tones between two multi-function dissolve units, with each tone serving as a signal to activate a different function.

These units are easy to program, but as you can see, the number of visual effects that you can produce is limited.

One of the early methods of controlling multiple projectors was with tone-control programmers.

Next in order of sophistication come **punched-tape programmers** that use either programming logic or electronic circuitry to create visual effects. Despite their different approaches to creating effects, both types of punched-tape programmers work basically the same. Each hole punched in a tape represents a function command to a dissolve unit. Each combination of holes represents an effect on the screen. So by changing the combinations of punched holes, you change your effects.

The major advantage of punched-tape programmers is that they give the experienced producer a ready means of checking a program. He or she can just "read" the program tape to quickly find mistakes.

The major drawbacks of punched tape are two: It takes longer to program (you may have to fill in or add holes to correct mistakes; this also makes experimentation more difficult). And tape readers are noisy when operating. (Their chattering sound can easily disturb an audience, especially when used for front-projection presentations.)

The most sophisticated programmers on the market today use **electronic microprocessors.** These units create—and store—their commands as digital information; each "bit" of information represents a command to one or more multifunction dissolve controls.

Electronic programmers are easy to program (after an initial learning period needed to become familiar with the programmer's logic). They also make it easier to correct mistakes. You can just erase the original instructions and replace them with new ones. And these programmers make it easier to experiment with different

combinations of visual effects. (Most programmers automatically move trays backwards or forwards as you add or delete cues.) You can also create an effect, then store it while you try another approach. This process can be continued until you create the effect you want. Then that effect can be placed in the unit's program memory, while the unwanted ones are simply erased.

However, these programmers also have several drawbacks. One is that you can't see your program as you can with paper tape (unless, of course, you have some system of linking your programmer to a standard computer printer or a data display unit). To check your programming, you have to step through the program cue by cue, sorting out preset commands from activating commands.

You can also lose your programmer's memory—and the time you put into developing it—if your electrical power is interrupted (although most manufacturers provide backup protection).

Finally, it's often hard to discriminate between a faulty set of programming instructions and a malfunctioning piece of equipment. In fact, unless you're an electronics expert, you may spend hours trying to find the source of a problem.

As you consider the three major types of programmers, remember to look beyond your current needs. Many beginning multi-image producers have bought the most basic and least expensive programmer on the market because it was exactly what they needed at the time. Within a year, however, as their own experience increased, they found that they wanted to use more projectors and more effects. The only way they could do this, of course, was to buy new equipment.

We're not suggesting you buy features and sophistication you'll never use. We **are** recommending, however, that you give yourself room to grow. Buy features you'll be able to use two and three years from now.

The Times Of A Producer's Life

If a producer tells you she's working in real time, don't think she's just stepped out of a time warp.

And, by the same token, if another producer tells you he's working in leisure time, don't think he's moonlighting or has made his work his avocation.

Real time and leisure time are terms used to indicate how a programmer's cues are created and stored.

Real time means actual time—the time measured by your watch. When a producer is programming in real time, he or she is creating the program second by second, without interruption, from the beginning to the end of a sequence or an entire presentation. Most real-time programming is done with a tone-control programmer and a tape recorder. As the recorder plays a presentation's sound track, the producer lays down audio tones on an adjacent track. The only way a producer can do this is to run the audiotape at playback speed—a real-time activity.

Leisure time in programming means nonreal time—or artificially manipulated time, or stored time. This is a difficult concept to grasp, so an example may make it more clear. If a producer were

programming a sequence in real time that consisted of a baseball spinning across a screen, he or she would have to create dozens of quick cuts with $1/10$-second pauses between each cut (less time than the blink of an eye). This is impossible using real-time programming techniques. Neither the person nor the equipment is equal to the task.

If the same producer were programming this sequence in leisure time, however, he or she could program the first quick cut, then a $1/10$-second pause, then the next cut, then another $1/10$-second pause, and so on until the end of the sequence. In other words, the producer wouldn't be tied to the demands of real time—the demands of a constantly ticking watch. Because the producer is storing the instructions—both those for dissolve/projector functions and those for the timing of those functions—he or she can work "at leisure." The producer is in control of time, rather than having time control the producer.

Later, when the program is played back, the visual effects will appear on the screen as if programmed in real time. The programmer will have converted its stored-time instructions into real-time action.

After you've selected the type of programmer you want and have isolated some of the product features you think you'll need, your next step is to try several of them. In the automotive industry, this is called "test driving," or getting a feel for the product.

The best way to test a programmer based on your needs is to call a manufacturer or dealer and ask for a demonstration. (For the names of manufacturers, check the Appendix at the end of this book or consult the annual **NAVA Audio-Visual Equipment Directory.**) But don't make your decision solely on the basis of a sales representative's demonstration. Ask to use the programmer; see how long it takes you to grasp the technique of programming. Plan to spend at least half a day,

preferably a full day, working with the equipment. And when the sales representative leaves, note your impressions of the unit's advantages and disadvantages. Then repeat the process with several other makes of equipment.

If the process seems long and involved, remember that you're about to spend from several thousand dollars to more than ten thousand dollars, and that the equipment you select is the equipment you'll be working with for several years. So take your time. Consider your decision carefully and thoroughly.

You're selecting the programmer that's **best for you;** so make **your** needs and considerations count.

(bottom) The 832 Micro Programmer, manufactured by Arion Corporation, controls up to 32 projectors with an internal microcomputer.

Programming The Presentation

Once you've pulled your raw materials together—the sound track, the motion picture sequences, the slides, the strobes, the spotlights, the ideas for visual sequences and special effects—you or your producer must sit down at one of those space-age-looking control consoles and orchestrate your multi-image presentation. That's right, orchestrate. The term "programming"—a word borrowed from the computer industry—seems too static to be accurate. We're talking about moving images across a screen with all the grace and beauty of a symphony or all the action and frenzy of contemporary rock music. The degree to which you'll be able to orchestrate movement will depend upon the sophistication of your programmer and the amount of experience you've had at its keyboard.

Unless you've had some experience with or exposure to one of the types of programmers illustrated and described on the previous pages, you're probably still wondering where to start. Obviously, we can't provide a complete course of instructions on how to operate all the different makes and models of programmers. In fact, even if we could, it wouldn't be very useful. Like any other skill, you learn best by doing. So what we'll do instead is describe the effects that can be created with most programmers and indicate when these effects might be used. Then we'll discuss an approach to programming that should be effective regardless of the type of equipment you use.

To understand what a programmer does, let's look briefly at its more sophisticated cousin—the computer. Today computers perform a seemingly endless range of tasks, from monitoring the operations of huge oil refineries to controlling the flow of traffic through metropolitan areas to regulating the temperature and humidity in homes and offices. As a result, the average person tends to view the computer as verging on the miraculous. And there's abundant evidence in the world to support that view. The computer's wide range of applications makes it seem ubiquitous. And its extraordinary data-handling capacity gives it the mystique of omnipotence. Just give the computer a task — any task — and it can do it.

In truth, all a computer does is manipulate 0's and 1's at fantastic speeds. That's all! Everything a computer does is based on binary arithmetic — the addition and subtraction of 0's and 1's. Once you understand and appreciate this fact, the feats performed by computers, while still impressive, are no longer mysterious or miraculous.

The same is true of programmers. Programmers can turn a screen into an explosion of light; they can control dozens of projectors and give tens of thousands of commands; they can take a series of slides and create the illusion of an animated cartoon. As a result, we tend to see the programmer as an audiovisual miracle-maker.

But when you strip away the aura of magic, you find programmers perform only three functions:

- They turn projector lamps on and off.
- They control the cycling of projector trays.
- They provide a means of timing these two functions.

(Depending on the make of programmer, the actual electrical or electronic operation that controls these functions may be carried out by an ancillary piece of equipment not unlike a basic dissolve-control unit.)

That's it. No magic. No miracles. No mystique. The fact is, programmers are just electronic devices that do what you've always been able to do with slide projectors. Only they do it faster and with greater precision than has been possible in the past.

The Show Pro II, manufactured by Audio Visual Laboratories, Inc., is a 40-channel, paper-tape programmer designed to control up to 10 projectors.

Audio Visual Laboratories' Eagle represents a further advance in the state of programming art. It is the first non-dedicated digital computer capable of programming multi-image presentations.

193

Creating Effects

If a programmer performs only three basic functions, it's evident that the programming effects you create will have to be based on one or a combination of these functions. You will turn lamps on and off at faster or slower rates while advancing your slide trays. Or you can create an effect using a single series of slides without advancing the trays. And you will have your effects take place at a rapid pace, a moderate pace, a slow pace, or with mixed pacing.

To repeat, these are the basic functions your programmer will perform. Your ability to orchestrate — **program** — these functions will give your presentation its individuality and style.

In creating your programming you will work with a number of basic techniques:

The **Show Pro V**, another Audio Visual Laboratories product, is a computerized programming system capable of controlling up to 15 projectors.

Dissolves Or Fades

This is the most basic of audiovisual effects — **the fading of an image until it disappears completely.** There are two types of dissolves or fades. In the first, the image on the screen fades to black. This is the equivalent of "cleaning" the screen before creating a new series of effects. In the second, the fading image is replaced on the screen by an emerging image; one image seems to be dissolving into another.

The programmer/dissolve system creates this effect by controlling the length of time it takes to turn projector lamps on or off. Most equipment on the market today offers dissolve or fade rates from less than one second to more than 30 seconds. This broad range of dissolve or fade rates enables a producer to create a variety of visual moods and to vary the rhythm or pacing of the sequences. In a straightforward presentation of graphic material, for example, a producer can use a moderate dissolve or fade rate to create an impression of "one set of data evolving into another." If, on the other hand, a producer wants to create a dream-like sequence, he or she might use long, slow dissolves or fades that communicate a feeling of smooth, gradual transition.

Fast Cuts

This effect — also called a **hard cut** or a **chop** by some equipment manufacturers — is a variation of the dissolve. But whereas the dissolve is a gradual fading of an image, a fast cut is just what its name suggests — **an image is either removed or substituted instantaneously.**

In terms of programmer operation, a fast cut is simply a dissolve at the fastest rate possible with your equipment. You will most likely use this effect when your intent is to establish a lively visual pace (when your sound track calls for a rapid succession of images) or when you want to "multiply" a series of images to create an overall impression.

As an example of the first application, consider the pacing you might want to establish for your visuals if your sound track consisted of disco music. Your visuals would have to move as quickly as the beat of the music or your sound and images would seem out of synchronization.

An example of the second application can be seen in the opening of a presentation we produced for a sales meeting held in Los Angeles. Our intent in developing this sequence was to communicate an overall impression of Los Angeles to the audience (most of whom had never visited the area) and to show these people the sights and atmosphere of the city — in less than four minutes. We decided the best way to accomplish this was to multiply the impressions — to create a "feeling" for Los Angeles using hundreds of images. Naturally, to achieve this, we used as many quick cuts as possible.

Our specific approach actually combined both applications: We selected several segments of fast-paced music, each of which were used to underscore the moods of different areas in Los Angeles — Hollywood, the downtown business area, Disneyland, and so on; then we programmed the opening sequence to the pace established by the music.

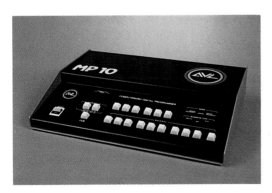

Audio Visual Laboratories' MP 10 is a computerized, 10-channel digital programmer with a solid-state memory.

Cut On/Dissolve Off.
Dissolve On/Cut Off

This effect combines the techniques of the first two. The effect created by quickly bringing up an image with a cut and then dissolving it off is analogous to an idea "popping into mind." We see an image instantly, then reflect on it as it slowly fades from our awareness. A sequence programmed in this manner visually simulates the process of thought: An idea suddenly comes into mind, is considered, then begins to fade until a new idea suddenly comes into mind, is considered, then begins to fade until. . . .

The opposite technique, to dissolve an image onto the screen and then cut it off, would simulate a different type of mental process. Here the images appear gradually, as if out of reverie, only to be interrupted and replaced by a sudden association with another image. So whereas the former technique seemed to simulate thought — one idea leading to another — the latter technique seems to simulate recall — one memory triggering another.

These analogies to the human thought process indicate some occasions when a producer might consider using one or the other of these techniques. If you're trying to get an audience to follow a train of thought, cut on/dissolve off may be an effective technique to use. If, on the other hand, you're trying to get an audience to form a conclusion by recalling experiences from the past (that are being stimulated by images on the screen), then the dissolve on/cut off technique may work more effectively.

Hold

This isn't an effect as much as it is the preparation for other effects. In a hold you **retain a slide (or slides) in the projector gate, with the lamp off,** because the slide is to be used again within a short period of time.

There are two reasons you would program a hold. You may want to use the same slide several times within a sequence. You could accomplish this by using duplicate slides; or you can hold the same slide in the projector. The advantage of the latter approach, of course, is that you free several slots in your slide tray for other slides, an important consideration in longer presentations. And more importantly, a slide that is held in the projector gate is ready immediately for reuse.

The second reason you would program a hold is to retain a series of slides to be used in flashing or an animation sequence. These effects will also be explained.

Clear Light's Star-3 programming system consists of a multi-screen programmer and multi-function dissolve units that also can be used in conjunction with one of two memory systems.

Wipes

Wipes are exactly what the name implies — **an image or series of images is removed from the screen in a gradual "sweep"** from left to right, right to left, top to bottom, or bottom to top. Wipes are created by programming a sequence of **dissolves** separated by a **time interval** (or **wait** or **cue link**). For example, if your screen contained three side-by-side images and you wanted them to change to three images of the same dimensions in a wipe from left to right, you would program a two-second dissolve on the left screen, followed by a half-second pause, followed by a two-second dissolve on the center screen, followed by a half-second pause, followed by a two-second dissolve on the right screen. When played back, the three dissolves would occur precisely, with half-second pauses between them.

Flashing

Flashing (also called **alternating**) is also exactly what its name indicates — you **program projector lamps to cut on and off in split-second intervals,** while **holding** slides in projector gates. Flashing is the programming equivalent of an exclamation point; it quickly calls attention to the images popping on and off the screen. We often use flashing to call attention to key words in a presentation (although only once or twice during a show; like any effect, flashing can become monotonous if used repeatedly). We've also used this technique to create an effect of flashing marquee lights around the borders of a screen.

Flashing With Lamps Getting Brighter Or Dimmer

This is a simple variation of the above technique. In addition to **holding** slides in the projector gates and programming the lamps to **flash,** you **program the lamps to grow brighter or dimmer.** When you program a gradual increase in lamp intensity, the effect created is that of movement within the projected image; the image seems to grow in size. The opposite technique — programming a gradual decrease in light intensity — makes the image seem to shrink in size as it fades from the screen.

Some producers use the terms **bumping** or **fade flash** to describe the gradual increasing or decreasing of lamp intensity.

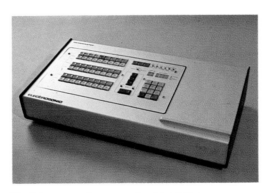

The Memomaster, manufactured by Electrosonic Systems Inc., controls up to 18 projectors.

Freeze

This term has two meanings. Some producers and equipment manufacturers use it to describe the technique of **projecting—or freezing—two images on the same screen area.** When used in this sense, the technique can also be called **superimposition.** The effects created by this technique are varied. They can be as simple as the projecting of key words or graphics over a pictorial background. Or the effects can be more complex, as when projecting two or more different pictorial images to create a multi-image montage in a single screen area. When the term is used this way, the effects created are called a **freeze on** (creating the superimposed image) and a **freeze off** (fading the superimposed image to black).

Other producers and manufacturers use freeze to describe a variation of lamp **bumping.** In this technique, the intensity of lamps is increased or decreased to a certain point and then frozen.

Electrosonic Systems' most sophisticated programmer is the Microcue Memory Programmer. It is capable of controlling up to 56 projectors.

Limited Animation

As we use the term, limited animation is an extension of **flashing,** only **holding** slides in a number of projectors (instead of just one or two) while projector lamps **cut on and off,** not simultaneously, but sequentially in a programmed cycle (or **loop**). Using this technique, you can cause an image to move across the screen or zoom up in size. And if the projectors have not been instructed to cycle (they can be instructed to cycle if you just want to use an image once), the effect can be repeated over and over again very quickly, because you're simply programming projector lamps to turn on and off. You don't have to be concerned with tray-cycling time, because the trays aren't moving.

Although we call this technique limited animation, it has unlimited applications. We've used it to spin a key word around an axis formed by its initial letter; to revolve a corporate logo and an identifying symbol in a series of reciprocating circles; to create the illusion of a baseball spinning across the screen.

Extended Animation

This is an extension of the above technique, only instead of holding a limited number of slides in projector gates, slide trays are advanced, introducing new slides into the sequence.

This technique can be used in much the same way animation is used on motion picture film. On one occasion, we used this technique to create the equivalent of an animated cartoon. The character in the sequence walked across the screen and in and out of doors. And by combining art with carefully masked slides, we were also able to move the character into pictorial images.

Special Effects

You can also use your programmer to create special effects not related to slide or motion picture projector operation. A programmer can control auxiliary functions—opening or closing a curtain, raising or lowering a screen, brightening or dimming house lights, activating special lighting effects, and so on. In fact, any device or appliance that can be activated or deactivated with a simple on/off switch can be controlled by a programmer.

No special programming technique is involved in creating these effects. All you need remember is that this function is available. When you use it and how you use it depend entirely on your imagination.

The Director-24, manufactured by Spindler & Sauppé, controls up to 24 projectors. It stores cues on a computer-grade Beta tape cassette.

The Products Of Programming

What sort of a presentation will you end up with?

Your presentation might be a totally canned show, in which case you'll be able to push one button when it's done, sit back and enjoy the show as a member of the audience. From a producer's point of view, this sort of presentation approaches nirvana. Everything is locked in the memory or on the cue track of your audiotape. Theoretically, everything will happen just as you planned and programmed it. The timing will be precise and flawless.

Or your presentation might be programmed to keep pace with a live narrator standing at a podium beside the screen. In this case, someone will have to follow the reading of the script, cueing the programmer at the appropriate points. This type of presentation will require more rehearsal time to perfect the timing. But there's no reason why this approach should be any less impressive or effective than a canned show. The information is all on the paper tape or in the computer memory of the programmer.

Yet a third possibility is that your presentation will be a combination of canned segments and live segments. This type of presentation will also require more extensive rehearsals to ensure smooth transitions between the two kinds of segments.

Programming Your Sequences

Now you're familiar with the most common programming techniques. But that doesn't qualify you to program a multi-image presentation. If anything, your reading qualifies you only to begin experimenting with the different techniques. Not until you've worked with each technique and observed the effects you can create with your equipment will you be ready to begin programming the sequences that will form an actual presentation. In short, you have to build a programming repertoire before you take the stage.

In the process of building your repertoire, you'll learn an interesting fact. **Programming isn't simply the mastery of technique.** Knowing how to create dissolves or flashing or animation doesn't make you a programmer, just a technician. To become a programmer, you must move beyond technique to art.

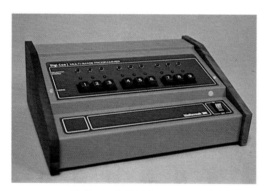

Wollensak / 3M Company's Digi-Cue programmer is capable of controlling nine functions using nine discrete audio tones.

To demonstrate the limitation of technique to yourself, program several sequences using a variety of techniques. Create one sequence using all dissolves, another sequence using all cuts, then use dissolve on/cut off for your third sequence, and reverse this for a fourth sequence. Finally, create some wipes using different dissolve rates and time intervals. In creating these sequences, plan on using the same number of slides for each technique.

Now run your program, **using the same set of slides for each technique.** (You may have to recue your slide trays after completing each technique.) Observe how the same slides begin to convey different subtleties and nuances of meaning, particularly affective meaning. Each technique used, each effect produced on the screen, will create a different emotional response within you. With an extended sequence of quick cuts, for instance, you may feel put upon, assaulted by information, or overwhelmed, and these reactions may lead you to resist the visuals, to adopt a "so what" attitude to the sequence. On the other hand, a sequence of slow wipes may draw you into the visuals, seduce you into paying greater attention, trigger memories that you will lay over the sequence, and thus strengthen its meaning to you. Affectively, you may feel an attraction, a willingness to participate imaginatively with the sequence, a receptiveness to its message.

That's what programming is all about—creating nuances of meaning. It's not merely throwing different patterns of images onto a screen using different programming techniques. **Programming is the use of techniques to create visual effects that strengthen the meaning of a sequence.** And this meaning is strengthened when members of the audience are **affected** (and not only informed) by the visuals.

When you begin to appreciate this fact, you'll realize two additional points:

First, programming isn't an activity you perform **after** the other aspects of production are completed. Instead, programming should be on your mind and in your planning from the moment you see—or write—the script.

Second, you don't start programming by asking, "What effect will I create now?" That's the way a technician works—putting technique on display. Instead, you begin programming a sequence by asking yourself, "What point is the script trying to communicate?" Push for a conceptual answer to that question. For example, the point of a sequence may be very specific: Sales at Company A grew 15 percent last year. The conceptual answer, however, is simply "growth."

Once you know what concept you're trying to communicate, you begin to plan your sequence, weaving both concrete and abstract into a whole. You do this by first deciding on the types of visuals you want to use to communicate the point of the sequence. Here the emphasis is on the specific: In our example, what specific visual subjects suggest "sales growth of 15 percent"?

Then you plan the **probable** programming for the sequence. Here the emphasis is on the abstract: Again using our example, what effect or combination of effects would suggest "growth"? Remember, at this point all you want is a feeling for the types of effects you may use; you don't need a detailed plan.

Finally, plan your shooting in light of your visual format and your probable programming.

It's only after you've shot and sorted the slides that you actually sit down to determine the slide-by-slide programming for each sequence. When you reach this point, the most helpful tool you can use is a programming sheet.

Programming Sheets And Show Books

Programming sheets are used to record the step-by-step instructions to be followed in programming a presentation. (See illustration for an example of how instructions are recorded.) You can buy basic programming sheets at some audiovisual dealerships. You also can obtain them from some equipment manufacturers who have designed programming sheets specifically for use with their equipment. Or you can follow the practice of many professional producers and design special sheets to use with the specific projection setups and screen configurations required for each production.

The use of programming sheets results in several important benefits. First, they enable you to "see" where images are going to fall on the screen, how they are going to look, what banks of projectors will be used to project them, and the order in which they'll be made to appear. Second, they let you "test" visual effects on paper before you do any special mounting or sandwiching of slides, or before you invest in any special optical effects. Third, they provide you with an accurate record of where slides sent for duping or optical effects have to be replaced. Finally, they record the slot positions of slides after they've been dropped into trays.

Although an entire presentation could be "preprogrammed" on paper, most producers we interviewed tend to program individual sequences as they work their way through a script. Then, after all sequences have been worked out and proofed, the information is transcribed to a fresh set of programming sheets to create a show book. This show book becomes the "bible" for that presentation. It includes the script, identification of each slide, its location in a tray, and the action that brings the slide onto the screen.

Once you're satisfied with a sequence on paper, you can drop the slides in the trays and create your program. (If you're working with a digital programmer that has a memory, you can experiment easily with the timing of effects. You can arbitrarily pick a screen action—a cut, for instance—and program the whole sequence with it. When you've finished that sequence, play it back to see how it looks. If it looks good, and all the visuals seem to work together, you can then go back and adjust the rates for the various screen actions, as well as the pauses between them. Remember, of course, this applies principally to presentations programmed on digital programmers. If you're programming in real time onto audiotape or using a punched-tape programmer, experimentation becomes time-consuming. If you're working without a program memory, you should solidify your planning before beginning actual programming.)

Once you've worked out a sequence to your liking, you should transfer its programming to audiotape, beta tape, or whatever permanent (or semi-permanent) material your programmer will handle—saving it for later reinsertion into the programmer. This step protects the programming from accidental destruction, should electrical power to the programmer be interrupted.

If you're programming a presentation with a narrated sound track, you'll be able to lay down cues or digital information for each sequence as you go along. If, on the other hand, you're programming slides to a piece of music, you'll want to pace the screen actions to fit that music, using a variety of fast cuts and different dissolve rates. Keep in mind, however, that if your programmer has a computer memory, you can go back effortlessly and experiment with variations until you find the effect you're looking for.

Using Your Capability Effectively

The new generation of programmers is so versatile and offers the producer so many different effects that it might seem almost impossible to take advantage of them all. So don't try, at least not the first few times you program a show. Remember that you started your production with a communications objective. Use the programmer to achieve that objective. Take advantage of the benefits it offers you as a producer to maintain the proper flow and visual continuity that will communicate your message. It isn't going to do you or the audience any good if everyone praises your effects but misses your message.

We aren't saying you should produce a static, slow-paced presentation. Use any of the effects that your particular programmer offers, but use them to support the message. Use flashing images or animation sequences to emphasize points in your message or even to help illustrate an action. Use slow dissolves and wide or multi-screen panoramas when you want to give your audience time to absorb information either visually or aurally.

As you gain experience with your programmer, you'll discover ways to use more of its features effectively in your presentations. And remember too that the programmer is only a tool. **You** have to provide the creative input.

Storing And Recalling Digital Memory

There are three ways of storing and recalling the digital information that makes up your program memory:

1. You can record the memory on audiotape, adjacent to your presentation's sound track. As the sound track is played back, the programming information is relayed through the programmer to the appropriate projection equipment.

2. You can record the memory on a separate reel of audiotape (or, where equipment allows, on a beta tape cassette), then reinsert—or "dump"—the memory into the programmer prior to a presentation. During the presentation, you use 1,000 Hz trip tones to signal the programmer to use the instructions in its memory. These tones can be activated manually (as when following the narration of a live performer) or they can be recorded on a presentation's sound track.

3. You can combine both approaches, having some programming information recorded directly onto the sound track and some stored in programmer memory, where it will be activated by trip tones. You might use this approach when your presentation consists of both live and canned segments.

The advantages of recording programming information directly onto the sound track are two: First, you save yourself a step when you reach the presentation site; you don't have to feed the memory from an audiotape to the programmer. This procedure can take you anywhere from 30 seconds (with

programmers that use beta tape to record programs) to several minutes when programming is recorded on conventional audiotape. Second and more important, when the programming instructions are recorded on tape, they are locked into the sound track. Even if the playback speed on the tape deck should change during a presentation, your show would not go out of sync. Sound and picture commands are linked.

The advantage of recording trip tones onto audiotape is that it makes it easier to change portions of your programming. If you try to change a portion of programming recorded on a presentation's sound track, you must use considerable precision to isolate the exact cues you want to change. Finding those cues is often difficult; you must watch for projector changes, then isolate the sound or word on which the change occurred. Making the changes is also time-consuming. You usually have to cut the segment of tape from the sound track, record the programming changes onto that segment, and then replace the altered segment into the original tape. This is the only way to protect against accidental erasure of the programming preceding and following the altered segment.

When your program is stored in the programmer, on the other hand, you don't have to worry about changing the cues on the audiotape; they stay where they are. You only have to change the programming in your programmer, a relatively simple procedure. When the trip tone is played back, the new set of instructions is activated.

David Wynne

David Wynne:
A Multi-Image Showman

David Wynne considers himself an old-fashioned showman, an impresario of light and sound and special effects who believes in entertaining audiences, not impressing them with technological fireworks.

"There's too much carnival and not enough show in many multi-image presentations," says Wynne, head of Dallas-based Multimedia Entertainment Corporation of America. "Too many presentations just show off the medium and how many pieces of equipment you can have running at one time, rather than telling a story or creating effects that blend together into a coherent, unified whole."

For Wynne, the story—coherent, unified, and **entertaining**—has always been the most important element in a multi-image presentation. This was true even during his first years as a producer, when he worked with several technically inventive men who created much of the equipment they would use to create equally elaborate multi-image effects. It was during this period that Wynne learned how effectively technology could be used to emphasize a story line, drive home a point, and create an impression. But he never lost sight of the fact that the story itself was still paramount.

Most of the productions Wynne worked on in those years were for industrial clients. During this period, as Wynne grew in experience, he

was also growing in ambition. He soon realized he was more interested in satisfying his own creative aspirations than he was in his clients' business needs.

After developing new ideas for attractions at the Six Flags Amusement Parks and studying the exhibits, displays, and presentations at Disneyland in Los Angeles, Wynne decided to try his hand at producing multi-image presentations aimed at general audiences. That decision, he says, "changed my life."

Wynne's first major general-audience success was **New Mexico—The Enchanted Land,** a 20-minute presentation telling the story of the people, places, and customs of the New Mexico area. It was an 18-projector show, with no motion and few special effects. But despite its relative simplicity, it was, says Wynne, "the most difficult presentation I ever put together. I had to learn how to tell a story, not create just another multi-image travelogue."

Wynne followed **New Mexico** with a more sophisticated presentation for the Green Bay Packer Football Hall of Fame. Next came his most ambitious project to date, **Jubilee,** which Wynne describes as a "multi-faceted visual experience." (See page 13 for a description and illustrations of **Jubilee.**)

Wynne found the production of these presentations "like making a commercial movie. You have to be entertaining, you have to put your imagination to work. You can't rely on a client's guidelines, nor can you assume you have a captive audience, naturally interested in your presentation, as you might have with a show used in a sales meeting."

Although many of the theatrical effects used in **Jubilee** are extremely sophisticated, Wynne doesn't think in terms of effects when producing a show. Instead, he's thinking about the story he's attempting to tell and the tools he might use to tell that story more effectively.

"Too many multi-image presentations lack an overall design, a rationale for what's on the screen," he says. "Producers use flashing or they project a dozen faces on the screen and then cut them and project a dozen more, but they don't know **why** they used this or that effect. It was just something they could do, so they did it. That's not good production. That's not good showmanship. To create effects that don't work, that don't advance or emphasize your story, is just a waste of money."

And for Wynne the second cardinal sin of multi-image production—following technological tomfoolery—is to waste money. He spent almost 18 months formulating his plans for **Jubilee,** then transformed these plans into a proposal that he used to raise the nearly $500,000 needed to produce and publicize the show. All this money was raised from investors in New Orleans, none of whom Wynne knew before approaching them with his proposal. He convinced them to invest with a combination of natural enthusiasm and hard-headed financial justification. Creatively, **Jubilee** is a success already; financially, it will pay dividends to its investors within two to three years of its opening.

"There's more potential for financial success in general-admission multi-image presentations than there is with commercial films," says Wynne. "There are no print costs, no distribution costs because you're dealing with one market, but within that market you have a high turnover of customers. Millions of tourists come to New Orleans every year, and every one of them is a potential customer for **Jubilee.**"

To attract these visitors to **Jubilee,** Wynne has established a group sales department that focuses its efforts on educational institutions, hotels, and tourist agencies. Another sales effort focuses on attracting the general public. "We have a continual turnover," says Wynne. "Different people are always coming to us."

Wynne believes that a good deal of the success of **Jubilee** and other similar productions lies in the power of multi-image presentations.

Part of this power, he feels, comes from novelty. People are just more naturally interested in new forms of entertainment—if they're used effectively. And to Wynne, that means don't go overboard on special effects when using multi-image.

He advises producers "to make sure people take something with them other than an impression of techniques. Viewers should have a positive **understanding** of the story you're telling or the area you're depicting. If all they do is go away wondering, 'How did they do that?' then you didn't produce a successful show. You just showed off."

Another reason Wynne feels multi-image productions are attracting viewers is that "they are more dynamic visually and they create a greater impact than do movie films or television programs."

"With film or television, all the action is confined to a well-defined screen area," he says. "What's happening on that screen may be very exciting, but the screen itself, the frame for that action, is very static. That's not true with multi-image presentations. You can vary the shape of the screen, vary the configuration of the image areas on the screen, you can switch from slides to film or film to slides or slides to holographic images or light displays or laser projections. There are any number of things you can do to keep an audience alert and interested, just as long as you **use** technology and not make it the central element of your show."

The entrance to the Jubilee theatre, located on Jackson Square in the heart of New Orleans' famed French Quarter, leads viewers into a multi-image presentation of the city's heritage.

Even though producers can employ a formidable array of multi-image equipment and techniques to capture audiences' attention, Wynne feels the production challenge is still greater with multi-image than with film. "With film you can rely on moving images to carry you through a complicated explanation or a review of facts or statistics. But with multi-image you have to rely on your imagination. You have to keep the screen—and your story—alive, and the only way you can do that is with good design."

For Wynne that just means another challenge, and the greater the challenge, the greater the rewards.

"For what I'm doing," says Wynne, "multi-image production has no limits."

This section examines the activities you must undertake before and after putting on your presentation. It emphasizes the importance of rehearsals and the need for an organized approach to transporting equipment. It also contains two checklists to help you follow through on the numerous details involved in setting up your equipment and preparing for the show itself.

SECTION IV
The Presentation

THE REHEARSAL

Checking Equipment And Programming

Your presentation is complete. Slides, film, sound, programming—all the elements are produced. Screens and equipment are in place. You're ready for your audience. **Almost.**

That's right—almost. Production is over, but you can't relax yet. Before you show your presentation to an audience, you have one final activity to complete—the rehearsal.

The rehearsal is **your** chance to review and refine your presentation. It's your chance to make mistakes, to let problems crop up, and to solve these problems without the pressures of an audience watching your efforts.

But while the pressure of an audience is off, the pressure of professionalism should still be on. A rehearsal is only as fruitful as the effort that goes into it. It's not an activity that should be performed perfunctorily, a routine run-through you try to get out of the way at the last minute so you can pack and ship your equipment. The rehearsal should be planned and conducted with as much care as you gave to production. It is your proving ground—an activity that could mean the difference between success and failure for your presentation.

Selecting A Presentation Team

Before you begin actual rehearsals, you should pick your presentation team—those people who will help you set up and operate the presentation equipment, both during rehearsals and the presentation itself. It's important that you select people who can work from rehearsals through presentation, because **what you want most of all from your presentation team is continuity of experience.** The person who loads the trays onto the projectors during rehearsal should be the same person who handles this activity at the presentation. The same is true for sound equipment, projectors, and programmer. With this type of followthrough, everyone knows what he or she should be doing at each moment of the presentation, so the possibility of slipups is minimized. And you, as the producer, have one less set of problems to worry about.

In selecting members of the presentation team, it would be ideal if you could select people experienced with the equipment they'll be monitoring. For example, you might want to have your sound recording specialist, if he's a part of your organization, operating your tape recorder and monitoring your audio system. The programming equipment should be operated by the person who programmed the presentation. In this way, people familiar with the equipment and the problems that might develop will be right where you need them in case of a malfunction—where their troubleshooting experience can be put to work.

Your presentation team should also have a leader. At the initial presentation, this leader should be the producer, the person who has overseen the development of the entire presentation and who knows the problems and requirements of each element in it. The presentation team leader will be responsible for coordinating and directing the efforts of the team members — who could number as many as four or five for a major production. The leader has to be certain that every person knows what his or her tasks and responsibilities are — who does what, when it is done, and how it is done. This sort of organized teamwork is critical — throughout rehearsals and especially on presentation day.

The Rehearsal Site

Selecting the site to stage your rehearsals may be a decision that's out of your hands. The room or work area where you produced the show may be the only suitable site for rehearsing. If so, you must make the best of your conditions.

If you can move to another site for rehearsals, try to find one that duplicates, as closely as possible, the conditions of the presentation site itself. This means finding a room or auditorium where you can:

1. Use the same equipment you'll use at the presentation. This is extremely important; in fact, it is crucial. Don't use the equipment at a rehearsal site just because it's there. That may be convenient, but it's not very practical. The one element you want to rehearse more than any other is your equipment.

2. Use the screen or screens you plan to use at the presentation. If this isn't possible, make sure your screen size and format are proportional to the actual screen or screens.

3. Use the same projection distance you'll use at the presentation. If you can't, you may have to use one set of lenses at rehearsals and another set at the presentation. If this is the situation you find yourself in, you'll have no chance to check actual image registration and brightness until you arrive at the presentation site.

4. Use the same projection setup — front projection or rear projection — as you'll use at the presentation site. If this isn't possible, you'll have another element to check out at the presentation site. You can, to some extent, preview a rear-projection presentation using front-screen projection. Of course, at some point, you'll have to insert all the slides into the trays in reversed position so you can become accustomed to viewing the presentation from that perspective (unless, of course, you're using mirrors or mirror lenses).

The "Full-Dress" Rehearsal

Once your equipment and programming are working to your satisfaction, your next step is the "full-dress" rehearsal. Now you're not as concerned with the functioning of the various elements of the show as you are with their **timing.** Like the director of a stage play, you want your performers and effects to work with clock-like precision.

During this full-dress rehearsal you should add one more element to the rehearsal site—an audience. This audience might consist of people from your own organization who haven't seen the show, or people from your client's organization.

The rehearsal of equipment, program and performers is critical to a successful presentation.

Their function is to watch—and constructively criticize—your presentation. They are surrogates for your eventual audience, so listen to their comments. You may not agree with them, but you should try to learn from them.

And if you learn something that seems valid, consider making changes to overcome the criticism. Remember, that's the point of rehearsals—to correct problems, not to have an audience applaud your work.

A rehearsal in front of an audience serves an additional purpose. It produces a semblance of "stage fright" in your presentation team. You, your narrator (if you're using a live narrator) and the people operating and monitoring the presentation equipment get to experience some of the nervousness you'll all feel on presentation day.

It's important that everyone involved becomes accustomed to working under the pressure of an audience. This nervous tension is going to be present on the day of the show and it has to be dealt with. Rehearsals won't eliminate this tension (although they may help reduce it somewhat), but they will teach your team members to function successfully despite the tension. And that, as every actor and speaker knows, is the hallmark of a professional performer. Once this full-scale rehearsal is out of the way, do it again (although you won't need an audience this time). And still again if possible. And when you get to the presentation site, **rehearse the presentation at least one more time.** The importance of rehearsals can't be overemphasized. As the time-honored adage states, practice makes perfect. So rehearse your presentation until you're happy and comfortable with the performance. Since producers are their own roughest critics, this do-it-until-it's-right attitude should assure you of a successful presentation.

210

TRANSPORTING YOUR PRESENTATION

Shipping

Unless your presentation is to be shown in your own community, you face the problem of getting it "from here to there." And that's a problem for a specialist.

Moving equipment cross-country—and making sure it's where it's supposed to be when it's supposed to be there—requires knowledge of carrier schedules and rates, packing requirements, insurance procedures, and pickup and delivery services. And unless you've worked with a shipping organization, you probably don't have the knowledge or experience to organize these details. Nor do you want to risk a lost or damaged shipment of equipment while trying to learn the procedures of the transportation industry.

So leave the problems of shipping your equipment to someone familiar with the business. That person may be someone from one of the various freight delivery organizations throughout the country or someone from the distribution department of your own organization. Whomever you choose, tell them what, when, and where you want to make shipment and when you must have delivery. Then leave the details to that person and concentrate your efforts where you can be most effective.

Packing. Make certain you have suitable shipping cases and that your equipment—**and spares**—are packed to withstand the jolts of a cross-country trip (and that applies even if you're shipping your equipment to the next town).

Many companies manufacture shipping crates and cases especially designed for the equipment used in multi-image presentations. These shipping cases come with special frames that hold projectors and dissolve modules. When you unpack these units, you merely lift out and set up the frames; projector adjustments are made within the frames. Almost all leading programmer manufacturers sell special shipping cases for their equipment to reduce the possibility of damage. The names of manufacturers are listed in the appendix.

If special cases aren't available for your equipment, you can build your own or have cases built to your specifications. The presentation department of Eastman Kodak Company had a case manufacturer custom build simple but effective shipping cases for its equipment. The cases were filled with the foam rubber, from which special pockets were cut for the various units of equipment, a variety of lenses, dissolve modules, power cords, junction boxes, even spare projection lamps. The principle used to design these units was "a place for everything and everything in its place." You can use the same principle to design and construct rugged shipping cases that meet your special requirements.

You can make your packing job easier if you label everything you plan to take with you—projectors, trays, screens, tape decks, amplifiers, loudspeakers, even the cables and power cords you'll need.

One of the easiest and most helpful ways to label your equipment is to assign some simple alphanumeric code to each major piece of equipment. Then ancillary equipment can be labeled by extending the code with additional numerals or perhaps a descriptive word.

For example, you might assign the letter "A" to all slide projectors, then number each projector you bring with you—A1, A2, A3, and so on. The trays for each slide projector could then be labeled with the addition of an extra numeral. A1-1, for instance, would indicate the first tray of slides to be used in projector A1. If you were putting on a major presentation, you might have trays labeled A1-2, A1-3, and so on; the dissolve module for your first bank of projectors could be labeled "A1-4 dissolve" (meaning the dissolve unit for projectors A1 through A4). The next dissolve unit would be labeled "A5-8 dissolve." Even the power cords for a projector or dissolve module could be labeled in this way—"A1-4 dissolve/cord."

Specially constructed shipping cases simplify the transportation of equipment and supplies.

And Don't Forget Your Toolbox

Richard III was willing to trade his kingdom for a horse.

And we've seen a number of multi-image producers who, on the day of a presentation, would have been willing to trade house and family for a screwdriver. Lacking someone with whom to trade, the frustrated producers used the edge of a dime or a soon-broken thumbnail to remove a resistant screw.

That's no way to be professional. Instead, you should buy a small toolbox (even a small fishing-tackle box will do) and stock it with the tools, gadgets, and supplies you're most likely to need when setting up or tearing down your equipment.

Some of the most obvious items to include are:

- screwdrivers
- nut drivers
- standard pair of pliers
- needle-nose pliers
- standard flashlight (with two D-cell batteries)
- penlight
- soldering iron, solder, and flux
- tweezers (to retrieve slides from gates or trays)
- skinny pencil (to poke slides down in gates if hung up)
- gaffer's tape
- masking tape (to label trays)
- electrical tape
- fuses for audio and projection equipment
- tape deck head demagnetizer
- motion picture film leader (opaque)
- splicers and splicing tapes (for movie film and audiotape)
- felt-tip markers
- grease pencils
- pocket-size notebook
- lens tissue and cleaner
- tape deck head cleaner, lubricant, and cotton swabs
- 3- to 2-prong electrical adapters
- audio adapter plugs and cords (e.g., RCA, ¼-inch phone, etc)
- extra cord for loudspeakers

That's just a starter list. If you think you'll need some other tool or gadget, put it in the toolbox. You may never use it. But it's reassuring to know it's there.

The numbering system would be extended to other pieces of equipment; motion picture projectors, for example, might be labeled B1, B2, B3, and so on.

With this sort of labeling system, you can pack everything you need quickly. And even more important, you can quickly sort out your gear as you pull it from the shipping cartons. This alone will make your setup job easier.

By using this coding system, you also can keep a list of the equipment you have in each shipping container. If your presentation is scheduled to move to a number of locations, you could even **assign** different pieces of equipment to each trunk. A 5 x 7-inch (127 x 178 mm) card taped to the inside of the trunk lid could list, by code numbers, the pieces of equipment to be packed in the container.

Insuring your equipment. Your shipping expert will handle the details of buying insurance coverage. But you must supply the serial numbers and the replacement value for each unit of equipment you ship.

Shipping rental equipment. If you've rented screens or equipment that must be shipped to the presentation site **by the rental agency,** the details are out of your hands. The rental company will want to know when you need the equipment; then they'll schedule delivery—usually a day before the actual presentation. They assume responsibility for having the equipment in your possession on that day. If for your own peace of mind you want the equipment at the site a day or two before you need it, you'll have to pay for the extra days of rental.

Receiving the equipment. In most cases, your equipment will be shipped to an individual in the receiving department of the facility where the presentation will be shown. You should make arrangements with this person during the production phase of your presentation.

This individual will verify the receipt of your equipment. But he or she won't open your cases to check the condition of your equipment. It's prudent, therefore, to check your equipment at your first opportunity. The facility's receiving department will also provide short-term storage and security for your equipment until you pick it up.

Having stated this, we should point out that there are plenty of exceptions to this sort of cooperation. Some hotels and convention halls don't have adequate storage space, so they don't make any provisions for receiving and storing equipment. The result is hair-raising for producers, to say the least. Equipment arriving at the site is left standing on receiving docks or just piled up in basement hallways. Sometimes delivery is even refused.

The only solution to this problem is to know beforehand precisely how your equipment will be treated upon delivery. If the people at the site will receive and store your equipment, fine; your problem is solved. If they won't, you may have to be at the site before your equipment arrives, so you can handle receipt and storage details.

The best advice we can give in this area is **don't take anything for granted.** Plan this aspect of your presentation as thoroughly as you do the actual production.

Setting Up—A Checklist

Setting up your equipment for the actual presentation should be a mechanical operation—not in the sense that it should be performed unthinkingly, but in the sense that it should be done according to a specific routine.

The following checklist will help you establish this routine:

1. The Projection Booth

Is it both safe and secure? The projection booth must be sturdy enough to hold both your equipment and the people who will operate it. It also must be steady enough so members of your presentation team can move around the booth during the show without causing projector movement. Draping a booth used for front projection not only improves its appearance, it also provides security against people tampering with wires or equipment. If you want additional security, you'll have to assign someone to stay at the booth.

Are power supplies adequate and safe from tampering? As mentioned earlier, you should ask for an "uninterrupted" power supply—power dedicated to **your** equipment only. The power source should also be safe from accidental tampering. It's not unheard of for a presentation to come to a halt because someone has accidentally pulled a plug.

Are your projectors properly aligned? Remember, the more projectors you use, the greater the probability you'll project keystoned images.

Make certain the projectors are centered vertically and horizontally in relation to each screen area. If, for example, you're using a 10 x 30-foot screen with two side-by-side full-frame images, the lenses of the projectors aimed at those two screen areas should be 15 feet (4.6 m) apart and seven and a half feet (2.3 m) from the ends of the screen. They should also be located at a level five feet (1.5 m) above the bottom edge of the screen. Slight departures from these ideal positions will have no significant effect on the images. Substantial departures, on the other hand, will cause image keystoning or distortion and may make it impossible to superimpose images exactly. The problem is greater with shorter focal-length lenses.

2. Screens And Seating

Have you set up your screen so it's parallel to your projection booth?

Have you registered your images on the screen? A number of companies make special registration slides just for this purpose.

Has the seating been set up according to your plans? Check the placement of seats, especially along the far edges of the room and to the rear of the projection booth (if you're using front projection). These are areas where viewers are most likely to encounter problems with image distortion or interference.

3. Loudspeakers And Microphones

Have the loudspeakers been set up to provide balanced volume throughout the audience area (and are they in phase with each other)?

Have you set the volume level you'll use during the presentation? In setting the volume, remember that your audience will "absorb" sound. So setting the volume slightly higher is usually necessary. (Masking tape can be affixed next to the volume control and marked for volume level.)

4. The Programming Equipment

Have you run a check on your programmer? Most manufacturers specify a procedure for checking all the circuitry. Also, be certain that all of the cables from projectors, dissolve modules, and any other equipment to be controlled by your programmer are properly connected. (Since programmer memories can be erased by even a momentary loss of power, it's wise to invest in a memory-protection battery unit. These units are sold by the equipment manufacturers.)

5. Spotlights And House Lighting

If you plan to use spotlights, **are they set and ready to go? And are the controls convenient and easy to use?**

Are the controls for the house lights easily accessible? If not, have you arranged to have someone at the controls, ready to activate them on command?

6. An Intercom System

If you must talk with presentation team members during a performance, **have you arranged to use an intercom system?** An intercom system lets the team leader communicate with all people involved in a presentation without disturbing people sitting near the projection booth. Instructions and cues can be given to people in the projection booth and to individuals in other areas of the projection site, where they may be operating lights, curtains, or other special effects.

THE DAY OF
THE SHOW

The Final Check

This chapter offers a bit of advice you're almost certain to ignore. Your presentation is completed. You've rehearsed it, fine tuned it, rehearsed it again. Now it's the day of the show and all your equipment is set up and functioning. To borrow a bit of jargon from the space program, everything is "go." So the advice of this book is — relax!

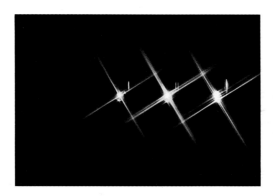

Impossible, you say. There are dozens of details to worry about. For instance, you're not sure if you repositioned the trays in the last bank of projectors. And you can't remember if you reconnected the number three dissolve module after you shifted the projectors around. And did you remember to tell so-and-so to tape down the wires behind the projection booth so no one inadvertently disconnects the power cords? And, what else did you want to remember to check?

So, instead of relaxing, you're beginning to grow tense as numerous questions race through your mind. And then there's the biggest question of all: Will everything work during the presentation as it did during rehearsals?

Granted, a book can't command you to relax. But it can help set your mind at ease — at least in terms of these last-minute questions and doubts. And the easiest way to do that is to divide your questions and doubts into two categories — one dealing with the details of your setup and the second with the possibility of equipment malfunction.

Let's take the latter of these two concerns first — primarily because it's easiest to deal with. The truth is the probability of equipment malfunction is slight. This isn't to say that equipment can't — and won't — malfunction during a presentation. It can and it will — but the occurrences are infrequent enough that the possibility shouldn't be a major concern on the day of the show.

These lingering doubts about equipment reliability — especially programmer reliability — are a carry-over from the early days of multi-image production. In those days — the early 1960s particularly — programmers weren't that reliable. They were, if anything, temperamental. They missed cues or they triggered prematurely or they refused to work at all. It was an era of potential and real horrors for multi-image producers, and their nightmarish memories of it still haunt them in the final hours before a presentation.

But product sophistication and reliability have increased dramatically since then, and the new generations of programmers—and dissolve units and projectors—are as reliable as almost any appliance in your house. Can they quit working? Sure. But so can your clock radio or your digital calculator. Yet you don't worry about the reliability of those products every time you use them—and that's the same attitude you should adopt with your audiovisual equipment.

With this category of worries eliminated, you can turn your concerns to the details of your projection setup. Here your questions and doubts are more valid. You do have dozens of details to think about, from the amount of life left in your projector lamps to the loading and connecting of projectors. Keeping all these details in your mind can tie your stomach into knots. So instead of worrying, what you need is a system for organizing and dealing with these details. The simplest system is a checklist.

Things to do one hour before the presentation:

There are probably two basically distinct feelings that can come over you about an hour before presentation time. The first is that if everything isn't right by now, it's probably too late. The second is that there must be at least a thousand things you should check, if you could only remember what they are.

In fact, the right feeling should be somewhere between those two. You **should** have prepared well for the presentation, so it **should** be ready to go. However, there are a lot of small details involved in making a show go off without a hitch, so it can't hurt to double check them.

The checklist below is designed to make you think about each of the details that **could** cause problems. To simplify the items on the checklist, we've broken them down into three major areas plus a miscellaneous category:

1. About the presentation:

• **Turn on the programmer, projectors, dissolve systems, audio equipment and any auxiliary equipment.**

This will serve two functions. With the first surge of power, any weak or overloaded circuits will trip a breaker. (Better now than an hour from now.) It will also force you to "proof" each piece of equipment to verify that it's operating.

• **Load and check your programmer memory.**

By now you're probably quite familiar with the procedure for your particular brand of programmer. Unfortunately, when we become familiar with a piece of equipment, we have a tendency to make assumptions.

• **Check to see that all projector dissolve modules and programmer controls are at proper settings.**

This, too, may seem obvious, but has frequently been the cause of false starts or worse. Controls are often changed after the final rehearsal and then not reset.

• **Check sound level from tape source, film and microphones.**

These levels might have been altered after rehearsal and you wouldn't want the show to begin with insufficient level.

• **Check to see that all the trays are on the proper projectors.**

Frequently, there's a need to clean that one last slide, or check its position in the mount just after the last rehearsal. As impossible as it might seem, the trays could have been put back on the wrong projectors.

• **Start the show.**

No, this isn't the real thing. It will, however, serve a very important function. If any projector lamps are going to blow, they'll most likely do it the first time they come on. You'll also get a chance to check focus one last time.

• **Reset the programmer memory.**

• **Check all tray positions.**

• **Check motion picture cue points.**

• **Check audiotape cue point.**

• **Check out intercom or walkie-talkie system.**

Too often, a common communications breakdown occurs because someone forgot to turn on their headset or walkie-talkie.

• **Check out auxiliary equipment.**

Special effects like spotlights, motorized drapes, screens, water displays, and so on, are generally out of sight and therefore out of mind. Make sure they're ready to go.

2. About the site:

• **Check the seating arrangements.**

Be sure no one has shifted chairs during the course of rehearsals or left coffee cups and full ashtrays.

• **Check all draperies around screen and projection booth.**

While draperies help give the presentation site a clean and uncluttered look, they often get in the way (and are moved) during rehearsals.

• **Notify facility management that the show is about to begin.**

They should already be aware of the fact, but it doesn't hurt to remind them. Also, it's a good idea to request that the house electrician be on hand or at least on call.

3. About the personnel:

• **Check all personnel for positions and cues.**

It won't hurt to coordinate cues for each person one last time, especially relating to house lights, spotlights and sound.

• **Do necessary personnel have copies of the script or agenda in hand?**

How many times have you discovered at the last moment that someone left his or her copy of the script back in the room?

• **Check to see that appropriate personnel have their flashlights.**

Like scripts, flashlights have a way of being left somewhere else. If you have a tray change, or if motion picture sequences have to be visually cued, a misplaced flashlight can be disastrous.

• **Do all staff members know what to do if . . . ?**

This parallels having fire drills at home. We don't like to consider the need, but we'd long regret not doing it if the situation arises.

4. Miscellaneous:

• **Are the entrances to the presentation site staffed?**

This doesn't seem like a responsibility that you, as the producer, should have to bear, but it could prove embarrassing to your client or your management if no one greeted the people arriving for the presentation and directed them to the seating according to schedule.

• **Check agenda and starting time with client.**

How many times have you found out that the client or a management representative has moved the starting time up or back fifteen minutes and forgotten to let you know? And how often have they decided to add one more activity to the agenda — and it's to take place immediately before the presentation is scheduled to start?

If you've gone through all the items on the checklist (and any other items that are unique to your presentation), you should be ready. In addition, you should have consumed the better part of the last hour before showtime, which leaves you with very little time to stand around trying to calm your nerves.

What Happens If . . . ?

Your presentation is under way. In almost all cases, it will go on as planned. All you need to do now is enjoy the applause and accept the congratulations. Your presentation is a success.

How ideal it would be if this book could end here! But . . .

Every once in a while you're going to put on a presentation where something goes wrong. The problem may be minor, a mere annoyance, or it may be major, a certified, this-can't-be-happening-to-me disaster. In either case, when these problems pop up, your only recourse is to be prepared. And the best way to prepare yourself for problems is to anticipate them and train yourself to deal with them.

Here are the most common problems you're likely to encounter:

1. What happens if a lamp burns out?

Of all the problems you'll face, this is the most commonly feared. For the most part, it's also one you can avoid by keeping a log of the hours a lamp has been used. The simplest way to do this is to affix a strip of masking tape under or on the side of each projector. Then, each time the projector is used, write the date and the number of hours the lamp was in use. When you approach the manufacturer's estimate of useful lamp life (see chart at left), replace the lamp. You'll have solved the problem before it develops. If a lamp does burn out, there are two quick ways to remedy the problem.

One. With the power still on, note the slide number at which the tray is set and remove it. Then replace the projector with a spare one. Plug the new projector into the dissolve control power module, and replace and resynchronize the slide tray.

Two. Replace the lamp. To do this, **you must turn off the projector.** Next, using a glove or thick cloth, remove the burned-out lamp and replace it with a fresh one. Then replace and resynchronize (if necessary) the slide tray. (Both of the above methods will probably require projector realignment.)

Neither of these solutions is necessarily easier or better than the other. The one you use will depend on your particular situation. If your projectors are set up so the lamps are easily accessible, you're probably better off replacing a lamp. On the other hand, if you have to pull out a projector to replace a lamp, you'll solve the problem sooner by simply replacing the projector. Then you can replace the lamp with less worry.

Lamps for **Kodak Ektagraphic** Slide Projectors

Projector Model	Lamp Code	Watts	% Relative Brightness		Life-Hours	
			High	Low	High	Low
AF-1, AF-2, AF-3, B-2, B-2AR	ELH	300	100	70	35	105
	ENG	300	130	90	15	50
	ENH	250	65	50	175	330
E-2, AF-2K	ELH	300	—	65	—	105
	ENG	300	—	85	—	50
	ENH	250	—	45	—	330
S-AV2000	EHJ	250	—	—	50	200

Don't Keep The Audience In The Dark

When you must stop your presentation to solve a problem, be sure to remember your audience. Don't keep them in the dark—literally or figuratively.

When something goes wrong, your narrator or presentation "host" should read a prepared statement to the audience, something along the lines of:

"We appear to be having some technical problems. Someone is coming up with information now, and I'll pass it along to you as soon as I get it."

Reading a statement such as this is a sign of professionalism. It tells your audience you have matters under control; it also encourages them to wait patiently while you solve the problem.

You promised the audience additional information, so be sure you get it to them. Once the producer—or the person in charge of the presentation—determines the nature of the problem, he or she should relay to the speaker specific information on what went wrong and approximately how long it will take to correct the problem.

If the repair can be made quickly, the presentation spokesperson should ask the audience to remain seated. If the problem is more serious, you'll have to choose your best course of action. You can:

1. Invite the audience back for a later showing (if indeed the problem can be solved by then).

2. Reschedule the showing for a later date.

3. Offer the audience a full refund (if you've charged admission) or a raincheck for a later performance.

Above all else, act professionally. Don't panic. Don't run around in a frenzy trying to solve the problem. Nothing destroys the confidence of an audience more than the sight of a producer who doesn't seem to know what he or she is doing.

The technique you'll use to resync your projector depends on the programming equipment you're using. With some units, the programmer itself will advance the slide tray to its proper position once the projector is turned on again. Other programmers indicate current tray position on their digital displays. You have to advance the tray to the corresponding slide number. With still other units you have to use your showbook to gauge the status of the presentation, after which you must position the slide tray to the next cue point. When the presentation reaches that cue point, you turn the projector on.

There's a third way to deal with the problem of lamp burnout—although it can't be classified as a remedy. It's more of a preventive measure, one used primarily with permanent presentations at exhibits, displays, and museums. This approach requires the use of a piece of equipment called an automatic lamper—a unit that removes and replaces a burned-out lamp in about three seconds. The unit, which attaches to a projector, is more expensive than commonly used projectors, so its use is usually limited to presentations that aren't continually monitored.

2. What happens if the programmer goes out of sync?

If you've hooked up your programmer and equipment properly, there's no reason for your presentation to go out of sync. Yet it sometimes happens. A stray, high-frequency radio or electronic signal may trigger the programmer, or it may be a sudden surge of electricity that does it. Whatever the cause, you have a problem.

Here again, the remedy you use to solve this problem depends to a large extent on the type of programming equipment you're using. But while specific techniques may differ, the procedure is basically the same.

One. In your showbook, find a reasonably close cue point—one that you can rotate your slide trays to **before** the narrator or your tape reaches it.

Two. With the sound track (or narrator) still going, disconnect your programmer memory. (Check your instruction manual for specific directions.) Recue the trays to the positions noted in the showbook. Also check for the projector lamp mode. Then, when the tape (or narrator) reaches the cue point, reconnect the programmer memory. Your show will be back in sync.

On some occasions, especially if your presentation is badly out of sync, you may have to stop the presentation itself to carry out this procedure.

3. What happens if the audio system goes?

When the audio system goes, you haven't got much choice. You're going to have to shut down the presentation and fix your equipment—if you can.

The most common problems in this area are blown fuses, broken audiotape, a loose connection between audio components, or a broken wire between components. To correct these problems, you'll need to use the toolbox consisting of the items listed on page 212. These tools and materials should enable you to repair common audio breakdowns. If the problem is internal—a short circuit or malfunction within one of the audio components—you'll have to replace the unit. Unless you have some spare components with you, this usually requires you to postpone the current performance.

Practicing For When Things Go Wrong

Don't wait until a lamp burns out or a programmer goes out of sync to learn how to solve these problems. Do what every organization does to prepare for an emergency—conduct "fire drills."

Practice removing and replacing projector lamps. First practice in full light, then in darkness simulating a presentation setting. When you get proficient, practice against a clock. Then practice during early rehearsals. By this time you should be able to change a lamp under pressure conditions.

You should practice resynching your program in a similar way. Learn how to disconnect the memory on your programmer so you can manually advance your slide trays. Then practice the procedure using five projectors, ten, or fifteen. Then, during rehearsals, deliberately throw your presentation out of sync and follow the procedures for resynching the slides.

By practicing these fire drills, you'll be able to solve the problems you're most likely to encounter during a presentation.

222

A Final Note

With your presentation over, you're going to be tempted to just throw your equipment and supplies into the nearest shipping case and go out and celebrate. Of course, with all the work you've done, you've earned the right. But don't give in to the temptation. To do so will be to invite problems the next time your presentation is to be shown.

Instead, when it comes time to tear down and pack up your equipment, do so with all the care you gave to packing prior to the presentation. The same rule applies: a place for everything and everything in its place.

When you follow this rule, you know you'll be ready to go the next time the presentation must be shown—whether that's two days, two weeks, or two months away. Programmer, power modules, projectors, lamps, cords, trays, slides, audiotape, film, script—all the elements of your presentation will be properly labeled and right where they're supposed to be. There'll be no last-minute scurrying around to rent an extra lens or buy another extension cord. You'll be prepared.

Don't take this advice too lightly. A lot of producers have said, "That may happen to others, but it'll never happen to me," only to go into near panic when they find a critical element of their presentation missing.

Maintenance

In addition to packing your equipment properly, you should perform necessary equipment and material maintenance as you put each item away—especially if the presentation is headed for another performance. If need be, clean and demagnetize your tape recorder heads, repair or replace frayed or loose loudspeaker cords, and replace projector lamps. The instructions for all of the equipment you're using also carry manufacturers' recommendations for periodic maintenance.

Make a list of those maintenance activities—similar to the ones auto manufacturers provide with new cars. Then, when the designated interval for service arrives, see to it that the maintenance is performed.

You should also check your slides and movie film for dirt and wear, and your audiotape for "dropouts" (sudden volume losses), torn splices, or any increase in hiss level.

Of course, under normal conditions, glass-mounted 2 x 2-inch slides will stay fairly clean and maintain their color for a reasonably long period. Under display or exhibit conditions, when the slides are being projected continuously for seven or eight hours a day, you'll have to replace them with a new set of dupes rather frequently.

Check your movie film for torn sprocket holes. If necessary, replace it with another print. And after every use of the film, wipe it clean with a soft cloth, using film cleaner with lubricant.

If you perform all of these maintenance checks, your presentation will be ready to go at all times. And it will look and sound as it did the first time it was shown.

Now It's Up To You

This book has covered a lot of material—from the factors you should consider when deciding to produce a multi-image presentation to the steps you should take to ship your production home after it's been shown. If you've read it carefully, you should have a thorough understanding of the decisions, considerations, and activities you must deal with during the course of a multi-image production.

But what you don't have—and what no book can give you—is **experience.** And until you have experience, you're not a multi-image producer. **So now it's up to you.** Go out and use the ideas, techniques, methods, and suggestions contained in this book. Convert your knowledge into know-how.

In short, **do.**

Produce a multi-image presentation. If you're a beginner, start on a small scale. If you have experience, introduce greater sophistication into your work. And if you're a professional, stretch your imagination to the limits of your equipment.

In doing this, you'll be writing the next—and most important—chapter in your multi-image instruction: **the chapter that deals with you.**

APPENDICES

Kodak Resources

Kodak Ektagraphic
Slide Projectors

Kodak Ektagraphic slide projectors — models E-2, B-2, B-2AR, AF-1, AF-2, AF-2K, AF-3, and S-AV2000 — are designed to meet the demanding needs of the professional user and incorporate many special features for multi-image applications. It's important to note those features that have helped to make Ektagraphic projectors the standard of the medium.

1. Ektagraphic projectors have accurate, repeatable horizontal and vertical slide registration. This feature is particularly important in multi-image presentations, where slide-to-slide registration must be precise.

2. Ektagraphic projectors have a heavy-duty motor that results in cooler operation. The benefit to the professional user is extended motor and bearing life.

3. Ektagraphic projectors (except model S-AV2000) have a dark-screen shutter — a feature that's especially beneficial to multi-image producers. This device interrupts the light beam from the projection lamp when the gate of the projector is empty (no slide), thus eliminating the need for opaque (dark) slides when programming and presenting your show.

4. Ektagraphic projectors, models AF-1, AF-2, and AF-3 are equipped with an autofocus on/off switch. When the switch is turned on, the autofocus mechanism automatically focuses each slide, once the first one has been manually focused. With the switch off, the autofocus function is defeated. This feature is especially desirable if you're using autofocus projectors in multi-image presentations. The benefit to you is that your projectors won't be re-focusing after each cue.

5. Ektagraphic projectors use both a varistor and a capacitor to control electromagnetic interference. This varistor-capacitor circuit is important in two ways. First, it extends clutch-spring contact life, especially beneficial in fast-paced multi-image shows where cycling can occur as often as once a second. Also, this circuit greatly reduces projector-generated interference with audiotape players and programmers — a potential cause of audio "popping" and unwanted slide changes.

6. The condenser lens in Ektagraphic projectors is coated (except for the Ektagraphic projector, model E-2). This coating reduces internal reflectance, giving you approximately eight percent more light output.

7. Ektagraphic projectors are equipped with a built-in three-wire grounding power cord, as required by many industrial and institutional safety codes (except model S-AV2000, which has a detached three-wire grounding power cord).

Kodak Slide Transparency Films

Kodak manufactures a wide variety of 35 mm still films suitable for preparing slides. Of the many film types available, each has its own characteristics, not all of which are desirable for every application.

Familiarity with these characteristics is necessary because exposure and filter compensation can enable you to use films other than those normally specified for a particular situation. For your convenience, we have included a table of color slide transparency films, showing the basic properties of each film. (If you need additional information on these or any other Kodak films, contact your local photo dealer or call the Kodak Professional and Finishing Markets Division Sales Office in your area. The addresses and phone numbers of the offices are given at the back of this book.)

Kodak Slide Transparency Films

Kodak Color Reversal Film	Balanced for	Film Speed and Kodak Wratten Filter Number								Processing—Kodak Chemicals
		Daylight		Flash		Photolamps (3400 K)		Tungsten (3200 K)		Electronic Flash	
		Speed (ASA)	Filter	Bulb	Filter	Speed (ASA)	Filter	Speed (ASA)	Filter	Filter	
Kodachrome 25 (Daylight) [135—20 and 36 exp.]	Daylight, Electronic Flash, Blue Flash	25	None	Blue	None	8	80B	6	80A	None	By Kodak labs and by photofinishers. Sent to Kodak by dealers or direct by users with Kodak Mailers. Process K-14
Kodachrome 40 Film 5070 [135—36 exp. only]	Photolamps (3400 K)	25	85	Blue / ASA 25	85	40	None	32	82A	85 ASA 25	
Kodachrome 64 (Daylight) [135—20 and 36 exp.]	Daylight, Electronic Flash, Blue Flash	64	None	Blue	None	20	80B	16	80A	None	By Kodak, other labs, or users. Sent to Kodak by dealers or direct by users with Kodak Mailers. Process E-6
Ektachrome 64 (Daylight) [135—20 and 36 exp.]	Daylight, Electronic Flash, Blue Flash	64	None	Blue	None	20	80B	16	80A	None*	
Ektachrome 200 (Daylight) [135—20 and 36 exp.]	Daylight, Electronic Flash, Blue Flash	200	None	Blue	None	64	80B	50	80A	None*	
Ektachrome 400 (Daylight) [110—12 and 20 exp.; 120; 135—24 and 36 exp.]	Daylight, Electronic Flash, Blue Flash	400	None	Blue	None	125	80B	100	80A	None*	
Ektachrome 160 (Tungsten) [135—20 and 36 exp.]	Tungsten	100	85B	—	—	125	81A	160	None	—	
Ektachrome 64 Professional (Daylight) [120; 135—36 exp.]	Daylight, Electronic Flash, Blue Flash	64	None	Blue	None	20	80B	16	80A	None*	
Ektachrome 50 Professional (Tungsten) [120; 135—36 exp.]	3200 K Tungsten	40 (at 1/50 sec)	85B	—	—	40 (at 1/2 sec)	81A	50 (at 1/2 sec)	None	—	
Ektachrome 200 Professional (Daylight) [120; 135—36 exp.]	Daylight, Electronic Flash, Blue Flash	200	None	Blue	None	64	80B	50	80A	None*	
Ektachrome 160 Professional (Tungsten) [120; 135—36 exp.]	Tungsten	100	85B	—	—	125	81A	160	None		

*If results are consistently too blue, use a CC05Y or CC10Y filter; increase exposure 1/3 stop when using a CC10Y filter.

S-16

S-30

S-74

Kodak Publications

Kodak's Motion Picture and Audiovisual Markets Division produces a wide selection of fact-filled, up-to-date publications addressing many different audiovisual-related topics. Described below are just a few of our many publications that should be of particular interest to you if you're working or planning to work in the field of multi-image.

Kodak Projection Calculator and Seating Guide (Publication No. S-16)

This useful tool — practical for any projection setup you'll encounter in single- and multi-screen shows (front or rear projection) — features a three-part dial calculator for computing factors such as image/screen size, projection distance, frame size, and lens focal length. It also includes a detailed table for determining the seating capacities of rooms of different sizes and shapes. (Dimensions: 6 x 9¼ in.) **[$4.95 per copy]**

Planning and Producing Slide Programs (Publication No. S-30)

This highly popular Data Book provides you with a wealth of practical ideas for use in preparing effective training, promotional, advertising, and educational slide presentations. Among the numerous subjects you'll encounter in this comprehensive, heavily illustrated publication are selecting your film and equipment for photography, exposing your film, shooting your artwork, duping your slides, handling and storing your slides, and much more. The 72-page publication contains 125 easy-to-follow photos, 30 line drawings, and many tables and charts. It's one that really should be included in your collection. **[$4.00 per copy]**

Kodak Sourcebook — Kodak Ektagraphic Slide Projectors (Publication No. S-74)

This Data Book — also a very popular item — is a comprehensive collection of Kodak Ektagraphic slide projector technical data and creative-use ideas that should help you achieve your communication objectives more effectively in the realm of 2 x 2-inch slide projection. The 72-page publication has five major sections that cover preparing for projection, projection lamps and lenses, adapting the projector for special applications, using slide projectors in multi-image situations, and maintaining slides and slide trays. **[$4.75 per copy]**

SOUND: Magnetic Sound Recording for Motion Pictures (Publication No. S-75)

This Data Book examines the vital role played by the sound track of any motion picture, and will bring you up to date on today's techniques for achieving top-quality sound reproduction. Starting out with a fascinating historical account of the origins of sound motion pictures, the book then proceeds with the "how" and "why" of sound recording. Among the many topics examined are preparing for sound recording — single- and double-system applications; script preparation; proper microphone use; double-system synchronization techniques; responsibilities of the sound recordist; postproduction; references for suppliers of sound re-

228

cording equipment; a glossary of sound terms; and much more. This 60-page publication contains 58 easy-to-follow photos and numerous drawings and charts. It's really a necessity if you're into multi-image and need to know the latest in motion picture sound recording. **[$6.25 per copy]**

Basic 2 x 2-Inch Slide Packet
(Publication No. S-100)

This packet consists of a select group of Data Books, pamphlets, and periodicals that answer many of the **basic** needs of the 2 x 2-inch slide program designer and user. Just a few of the many topics covered are planning and producing slide programs, selecting and using Kodak Ektagraphic slide projectors and allied equipment, effective visual presentations, image legibility, slide copying, slide artwork techniques, projection cabinets, and sources of commercially available slides. Also included in the packet are a number of free catalogs and resource listings that cover other applicable Kodak publications, Kodak visual communication films and slide programs, our current audiovisual equipment lines, dealers in Kodak audiovisual equipment, etc.

(**Planning and Producing Slide Programs** [S-30], **Kodak Projection Calculator and Seating Guide** [S-16], and **Kodak Sourcebook — Kodak Ektagraphic Slide Projectors** [S-74] are all part of this packet — as well as the many additional publications covering the topics listed above.) **[$20 per packet]**

The World of Animation
(Publication No. S-35)

This book is filled with information on the creative techniques needed to develop an animated-cel sequence. You'll learn about the rough drawing and cleanup stages, the concept of cel layering, the importance of background art, and more.

You'll also find detailed plans for building an animation stand at a fraction of the cost for most stands on the market. For under $400, you can make a stand that includes a professional Oxberry animation disc with peg bars.

And you'll find detailed information about planning, budgeting, and applications for animation in television commercials, public relations films, training and education films, and other areas.

Additional information includes sources for equipment and materials, a helpful glossary of terms, an exciting account of how animation developed, what's being done in the field today, and what opportunities there are for you.

Over 150 photos and illustrations include historical photos of animated scenes and early animators, line drawings of cels and animated sequences. (Approximately 125 pages; 8½" x 11".) **[$7.95 per copy]**

Ordering Information

For your convenience, you may order any of the publications described above by using the self-mailer order form at the back of this book, which includes a coupon for one free copy of MP & AVMD **Publications Index** (S-4).

S-75

S-100

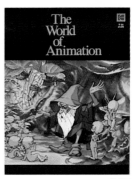

S-35

Equipment Manufacturers

(Inclusion of any suppliers referenced or illustrated in this book does not constitute an endorsement; there may be others completely satisfactory not listed or shown.)

Audio Equipment

Ampex
401 Broadway
Redwood City, CA 94063

Audio-Announcer
680 Bizzell Dr.
Lexington, KY 40504

Audiotronics
Box 3997
N. Hollywood, CA 91609

Califone International, Inc.
5922 Bowcroft St.
Los Angeles, CA 90016

Hitacho Sales Corp. of America
401 Artesia Blvd.
Compton, CA 90220

MacKenzie Laboratories
5507 Peck Rd.
Arcadia, CA 91006

Martel Electronics
970-A E. Orangethorpe Ave.
Anaheim, CA 92801

Otari
981 Industrial Rd.
San Carlos, CA 94070

Rangertone Research, Inc.
509 Madison Ave.
New York, NY 10022

Scully Recording Instruments
548 Kingsley Dr.
Los Angeles, CA 90020

Sharp Electronics
10 Keystone Pl.
Paramus, NJ 07652

Sony Corp. of America
9 W. 57th St.
New York, NY 10019

Studer Revox America, Inc.
1819 Broadway
Nashville, TN 37203

TEAC Corp. of America
7733 Telegraph Rd.
Montebello, CA 90640

Telex Communications
9600 Aldrich Ave. S.
Minneapolis, MN 55420

3M Co. Mincom Division
3M Center
St. Paul, MN 55101

Wollensak/3M Co.
3M Center, Bldg. 223-5E
St. Paul, MN 55101

Programmers and Dissolve Controls

Arion
825 Boone Ave. N.
Minneapolis, MN 55427

Audio-Sine
3415 48th Ave. N.
Minneapolis, MN 55429

Audio Visual Laboratories
500 Hillside Ave.
Atlantic Highlands, NJ 07716

AV Services, Inc.
2 W. 45th St.
New York, NY 10036

Avtek
120 High St.
Medfield, MA 02052

Clear Light Productions
PO Box 391
Newton, MA 02158

Columbia Scientific Industries
PO Box 9908
Austin, TX 78766

Commercial Electronics LTD
1335 Burrard St.
Vancouver, British Columbia V6Z 1Z7

Communications Control Co.
PO Box 707, 92652
1215 Dunning Dr.
Laguna Beach, CA 92651

EEG Enterprises
82 Rome St.
Farmingdale, NY 11735

Electrosonic Systems
4575 W. 77th St.
Minneapolis, MN 55435

Entré Electronics
PO Box 3122
Eugene, OR 97403

Kimchuk
34 Delmar Dr.
Brookfield, CT 06804

MacKenzie Laboratories
5507 Peck Rd.
Arcadia, CA 91006

Modern Media
10609 Quarterline Rd.
Apache Junction, AZ 85220

Optical Radiation
6352 N. Irwindale Ave.
Azusa, CA 91702

Optisonics HEC
1802 W. Grant Rd.
Tucson, AZ 85705

Pavco Electronics, Inc.
12810 Coit Rd.
Dallas, TX 75251

Spindler & Sauppé
13034 Saticoy St.
N. Hollywood, CA 91605

Technamics Co.
2232 Gardner Station
St. Louis, MO 63109

Tempo Audivision
290 Larkin St.
Buffalo, NY 14210

Tiffen Mfg.
90 Oser Ave.
Hauppauge, NY 11787

Trius Corp.
1169 Sonora Ct.
Sunnyvale, CA 94086

Wollensak/3M Co.
3M Center, Bldg. 223-5E
St. Paul, MN 55101

Projection Screens

Audio Visual Promotion Aids
466 Lexington Ave.
New York, NY 10017

Commercial Picture Equipment
(Div. of Bell & Howell)
5725 N. Broadway
Chicago, IL 60660

Da-Lite Screen
3100 State Rd. 15N.
Warsaw, IN 46580

Draper Screen
411 S. Pearl
Spiceland, IN 47385

Freen Screen
649 E. Lofstrand Ln.
Rockville, MD 20850

Knox Mfg. Co.
111 Spruce St.
Wood Dale, IL 60191

Spectro-Vue
8115 Commercial St.
La Mesa, CA 92041

3M Brand Polacoat Products
3M Company Visual Products Division
3M Center, Bldg. 220-10W
St. Paul, MN 55101

Trans-Lux News-Sign
625 Madison Ave.
New York, NY 10022

Wilcox-Lange
3925 N. Pulaski Rd.
Chicago, IL 60641

Projector Stands

American Professional Equipment Co.
2802 S. MacDill Ave.
Tampa, FL 33609

**Audio Visual Consultants
& Contractors**
6875 E. Evans
Denver, CO 80222

AV Technologies, Inc.
PO Box 10583
Dallas, TX 75207

Buhl Optical
1009 Beech Ave.
Pittsburgh, PA 15233

Chief Mfg.
PO Box 96
Savage, MN 55378

Columbia Scientific Industries
PO Box 9908
Austin, TX 78766

Communication for Business, Inc.
2525 Stemmons Expwy. #266
Dallas, TX 75207

Da-Lite Screen Co. Inc.
PO Box 629
1219 Winona Ave.
Warsaw, IN 46580

Kimchuk
34 Delmar Dr.
Brookfield, CT 06804

MacKenzie Laboratories
5507 Peck Rd.
Arcadia, CA 91006

Optisonics HEC
1802 W. Grant Rd., 101
Tucson, AZ 85705

Pran Audio Visual
1001 Broadway
New Braunfels, TX 78130

RMF Products
PO Box 413
Batavia, IL 60510

Tiffen Mfg.
90 Oser Ave.
Hauppauge, NY 11102

WTI Corp.
31423 Camino Capistrano
Suite 227
Laguna Niguel, CA 92677

Slide Projector Lenses

Buhl Optical
1009 Beach Ave.
Pittsburgh, PA 15233

**GAVI
General Audio-Visual**
306 Hempstead Ave.
Malverne, NY 11565

**NAVITAR
D.O. Industries, Inc.**
317 East Chestnut St.
East Rochester, NY 14445

16 mm Motion Picture Projectors

Arriflex Corp.
1 Westchester Plaza
Elmsford, NY 10523

Atlantic Audio Visual
630 9th Ave.
New York, NY 10036

Audio Visual Devices
38 Smith St.
Irvington, NJ 07111

**Bell & Howell/Audio Visual
Products Division**
7100 McCormick Rd.
Chicago, IL 60645

Bolex (USA)
250 Community Dr.
Great Neck, NY 11020

Carbons, Xetro Products Division
10 Saddle Rd.
Cedar Knolls, NJ 07927

Eastman Kodak Co.
343 State St.
Rochester, NY 14650

EIKI International
27882 Camino Capistrano
Laguna Niguel, CA 92677

Elmo Mfg. Corp.
70 New Hyde Park Rd.
New Hyde Park, NY 11040

Kalart Victor
Hultenius St.
Plainville, CT 06062

MacKenzie Laboratories
5507 Peck Rd.
Arcadia, CA 91006

Optical Radiation Corp.
6352 N. Irwindale Ave.
Azusa, CA 91702

Rangertone Research
509 Madison Ave.
New York, NY 10022

Singer Education Systems
3750 Monroe Ave.
Rochester, NY 14603

Strong Electric
87 City Park Ave.
Toledo, OH 43697

Viewlex Audio-Visual
PO Box 200
Holbrook, NY 11741

W. A. Palmer Films, Inc.
611 Howard St.
San Francisco, CA 94105

Slide Projectors

Atlantic Audio-Visual Corp.
630 9th Ave.
New York, NY 10036

AV Services, Inc.
2 W. 45th St.
New York, NY 10036

Behavioral Controls, Inc.
PO Box 480
1506 W. Pierce St.
Milwaukee, WI 53201

Bergen Expo Systems
1088 Main Ave.
Clifton, NJ 07011

Buhl Optical
1009 Beech Ave.
Pittsburgh, PA 15233

Eastman Kodak Co.
343 State St.
Rochester, NY 14650

EIKI International, Inc.
27882 Camino Capistrano
Laguna Niguel, CA 92677

GAVI General Audio-Visual, Inc.
306 Hempstead Ave.
Malverne, NY 11565

George R. Snell Associates, Inc.
155 U.S. Route 22 East
Springfield, NJ 07081

International Audio Visual, Inc.
15818 Arminta St.
Van Nuys, CA 91406

MacKenzie Laboratories
5507 Peck Rd.
Arcadia, CA 91006

Optical Radiation Corp.
6352 N. Irwindale Ave.
Azusa, CA 91702

Rangertone Research
509 Madison Ave.
New York, NY 10022

Singer Education Systems
3750 Monroe Ave.
Rochester, NY 14603

Sirtage
PO Box 30691
Umstead Industrial Park
Raleigh, NC 27612

Strong Electric
Johns-Manville Sales Corp.
PO Box 1003
87 City Park Ave.
Toledo, OH 43697

Slide Mounting Materials and Equipment

Byers Photo Equipment
6955 SW Sandburg St.
Portland, OR 97223

Ed-Tech Service
295 Main St.
Chatham, NJ 07928

Emde Products
2040 Stoner Ave.
Los Angeles, CA 90025

Eumig (USA)
Lake Success Business Park
225 Community Dr.
Great Neck, NY 11020

Heindl Masks 'N' Mounts
200 St. Paul St.
Rochester, NY 14604

H.P. Marketing
98 Commerce Rd.
Cedar Grove, NJ 07009

Kaiser Products
3555 N. Prospect St.
Colorado Springs, CO 80907

Karl Heitz
979 Third Ave.
New York, NY 10022

Pako
6300 Olson Memorial Hwy.
Minneapolis, MN 55440

Photo Plastic International
1639A 12th St.
Santa Monica, CA 90404

Wess Plastic
50 Schmitt Blvd.
Farmingdale, NY 11735

Zerro Photo Products
1327 2nd Ave.
New Hyde Park, NY 11040

Music Libraries

**(The BBC Sound Effects Library)
Films for the Humanities, Inc.**
PO Box 2053
Princeton, NJ 08540

Boosey & Hawkes
30 W. 57th St.
New York, NY 10019

Capitol Records, Inc.
1750 N. Vine St.
Hollywood, CA 90028

Chappelle Music Company
6255 W. Sunset Blvd.
Los Angeles, CA 90028

**De Wolfe Music Library &
De Wolfe Sound Effects Library**
25 W. 45th St.
New York, NY 10036

Emil Ascher, Inc.
666 Fifth Ave.
New York, NY 10019

Gary & Timmy Harris, Inc.
236 W. 55th St.
New York, NY 10019

General Music Corp.
6410 Willoughby Ave.
Los Angeles, CA 90038

Instant Music
5325 Sunset Blvd.
Los Angeles, CA 90027

MusiCues
1156 Avenue of the Americas
New York, NY 10036

Musifex, Inc.
45 W. 45th St.
New York, NY 10036

Regent Recorded Music
6255 W. Sunset Blvd.
Los Angeles, CA 90028

**Sam Fox Publishing
Company, Inc.**
62 Cooper Square
New York, NY 10003

Southern Music
Publishing Company
6922 Hollywood Blvd.
Los Angeles, CA 90028

Thomas J. Valentino, Inc.
151 W. 46th St.
New York, NY 10036

Animation and Copy Stands

Animation Sciences
114 Miller Pl.
Mt. Vernon, NY 10550

Arriflex Corporation
1 Westchester Plaza
Elmsford, NY 10523

Bencher
333 W. Lake St.
Chicago, IL 60606

E. Leitz, Inc.
Lind Dr.
Rockleigh, NJ 07647

Forox Corp.
393 West Ave.
Stamford, CT 06902

Impact Communications, Inc.
9202 Markville Dr.
Dallas, TX 75243

Matrix Division/Leedal, Inc.
2929 S. Halsted St.
Chicago, IL 60608

Maximilian Kerr Assoc.
2040 State Highway 35
Wall, NJ 07719

Oxberry Division of Richmark
Camera Service
180 Broad St.
Carlstadt, NJ 07072

Sickles
PO Box 3396
Scottsdale, AZ 85257

Slide/Art Production Services

Computer Image Corp.
2475 W. 2nd Ave.
Denver, CO 80223

Genigraphics

Center 1
General Electric Company
CSP-Building 3, Room 17
Syracuse, NY 13221

Center 3
General Electric Company
4929 Wilshire Blvd., Suite 960
Los Angeles, CA 90010

Center 5
General Electric Company
219 E. 42nd St.
New York, NY 10017

Center 7
General Electric Company
LNG Tower — 2919 Allen Parkway
Houston, TX 77019

Metacolor
855 Sansome St.
San Francisco, CA 94111

Visual Horizons
208 Westfall Rd.
Rochester, NY 14620

Wilson/Lund
1830 6th Ave.
Moline, IL 61265

Shipping Containers

Anvil Cases
4128 Temple City Blvd.
Rosemead, CA 91770

Bobadilla Cases
2302 E. 38th St.
Vernon, CA 90058

Cargo Case Division
Icom, Inc.
237 Cleveland Ave.
Columbus, OH 43215

Cases, Inc.
1745 W. 134th St.
Gardena, CA 90245

Chief Mfg. Co.
PO Box 96
Savage, MN 55378

Fiberbilt Photo Products Div.
Ikelheimer-Ernst, Inc.
601 W. 26th St.
New York, NY 10001

Glossary

Acoustical Balance — Fine tuning a sound system for optimum reproduction, taking into account the acoustical characteristics of the presentation site.

Alternate Flashing — An effect where the lamps in two projectors are alternately turning on and off, but out of phase with each other. While projector "A" is on, projector "B" is off. Flashing can also be done utilizing only one projector.

Answer Print — The first print (combining picture and sound, if a sound picture), in release form, offered by the laboratory to the producer for his/her acceptance. It is usually studied carefully to determine whether changes are required prior to printing the remainder of the motion picture prints.

Burn In — An effect which can be accomplished in-camera or on the screen where white lettering can be "burned in" to a darker image area. The technique is accomplished in-camera by photographing the scene (or duplicating another transparency) and double exposing a type slide made with high-contrast film so the letters fall in a dark area. For a burn on-screen, one projector presents an image while a second projector "burns in" white type into an area of the image.

Chop — A chop is similar to a cut in that as one projector lamp goes down, another comes up. With a chop, however, the down-going projector begins advancing or the shutter closes before the lamps (or slides) change, allowing the shutter of that projector to "chop" off the light. This causes a harder visual effect than does the standard cut. A chop may also be called a "hard cut" or "fast cut."

Composite Print — A positive movie film having both picture and sound track images on the same film, which may be in editorial or projection synchronization.

Constant-Illumination Presentations — Presentations where the screen is never dark from beginning to end. Changing scenes through the use of dissolves, cuts, chops, wipes, etc, eliminates the problem of a dark screen during slide changes.

Continuous-Tone Masks — Masks made on film with a clear base (such as Kodalith or Ektachrome films) and processed continuous tone. The edges are feathered in such a way that when sandwiched with the appropriate transparencies and properly aligned, they create full-screen effects with no telltale line showing where one portion of the image ends and another begins (also called seamless masks, soft-edge masks, or shadow masks).

Cue Link — Several cues can be "linked" together to form a chain of cues. When the first cue is executed, the remaining cues in the chain will also be executed in the sequence and time frame at which they were programmed. This is helpful in an animation sequence, for example, where it would be impractical to program each step using pulses from the audio track.

Cut — An instantaneous switch from one projector to another. As one projection lamp cuts on, the other cuts off. (So named because it approximates a motion picture cut — an instantaneous transition from one shot to another.)

Digital Information — Information stored in program memory or on magnetic tape that, when played back on the appropriate programmer (or playback unit), will cause the desired sequence of events to take place within the equipment under program control. Digital information is stored as +/− bits, rather than as continuously variable (analog) information.

Dissolve — An effect in which one scene gradually fades out as a second scene fades in. The dissolve effect is usually achieved by varying the intensity of the projection lamps in the two projectors involved. The time period necessary to complete the dissolve effect is called the dissolve rate. The dissolve effect is sometimes called a lap dissolve. (See Fade.)

Dissolve Bank — Two or more slide projectors (generally aimed at the same screen area) controlled by a single dissolve-control unit.

Double-System Recording — An arrangement whereby picture and sound are recorded on separate strips of film, usually with the provision that the two mechanisms, camera and recorder, will pass identical lengths of film in a given time (once both machines have come up to running speed) so picture and sound can be matched synchronously throughout their length in the editing process.

Fade — A gradual increase or decrease in projection-lamp intensity as in fade-in (image slowly comes up on dark screen) and fade-out (image slowly goes down, leaving screen black).

Field Guide — A sheet of heavy celluloid or acetate that is used for positioning artwork in the area or field recorded by the camera. When the guide — upon which all of the camera fields are indicated — is placed on the registration pegs on the platen area, there is no need to view through the camera. The guide shows the exact area that will be photographed.

Film Log — A step-by-step written record of the activities involved in making a motion picture. Separate logs are usually kept for camera and sound recording activities.

Folding — A projection technique utilizing mirrors so that the projector-to-screen distance can be shortened while retaining the same image size.

Gutters — The narrow dark bands separating images where two or more slides are projected at the same time without the use of continuous-tone masks.

Interlock — Any arrangement permitting the synchronous presentation of picture and matching sound from separate motion picture films. The simplest one consists of a mechanical link connecting projector and sound reproducer being driven by a common synchronous drive.

Jump Cut — A discontinuity of the action within a shot or between two shots due to removal of a portion of movie film, or due to poor pictorial continuity.

Keystoning — A geometrical image distortion resulting when a projected image strikes a plane surface at an angle other than perpendicular to the axis of throw, or when a plane surface is photographed at an angle other than perpendicular to the axis of the lens.

Leisure-Time Programming — Programming visual effects using a separate program memory (or paper tape) where the programmer is told in what order and at what speed to repeat those effects.

Lip Synchronization — The relationship of sound and picture that exists when the movements of speech are perceived to coincide with the sounds of speech (also called lip-sync).

Memory Dump — The capability of programming equipment to store program memory on magnetic tape.

Panorama — A full-screen image created by projecting three or more slides simultaneously. It can be a single, uninterrupted image covering the entire screen (see Continuous-Tone Masks) or a composite image made up of several separate and distinct portions.

Phasing — A problem that may be encountered when mixing stereophonic sound onto a monaural sound track. Instead of getting the total of both signals, in some cases one signal cancels out the other, thus causing a dropout. May also refer to problems created when projectors and control equipment are connected to out-of-phase power circuits, often resulting in erratic operation.

Posterization — An image with a flat, poster-like quality. High-contrast lith film is used to separate the continuous gray tones of the image into a few distinct shades of gray. These separations are then photographed using multiple-exposure techniques and colored gels to form a color-composite image.

237

Real-Time Programming — Programming visual effects directly onto the audiotape in the exact time frame (relative to the sound track) at which they must be repeated.

Release Print — A composite print made for general distribution and exhibition after the final trial composite, answer, or sample print has been approved. It is in projection synchronization.

Sandwiching — Placing more than one piece of slide film in the same slide mount to create a desired effect. (Used to mask a portion of the image; to lay down dark type in a light area of the image; to create double-exposure effects, etc.)

Sequential Loop — Used in programming to create cyclic animation effects. A series of images is projected more than one time without advancing slide trays.

Single-Frequency Electronic Cues — Electronic pulses at a given frequency that are recorded on the audio track at predetermined points. The pulses will "tell" a simple dissolve-control unit to cycle or a complex programmer to execute the next cue in its memory.

Single-System Recording — An arrangement for recording original picture and sound simultaneously on the same strip of film. Since sound and picture demand incompatible original film and processing characteristics for optimum fidelity in each case, a compromise is mandatory in photographic single-system practice. Usually, prestriped film is used in modern single-system cameras, with improved results. With picture and corresponding sound at different positions on the same film, however, unique editorial dilemmas are frequent in editing single-system films, unless the sound is first transferred to a separate recording material.

Slide Registration — Certain visual effects depend on accurate registration of the images for their impact. Care must be taken in creating the images (whether from artwork or live photography), mounting the slides, and aligning the projectors to ensure that the effect is optimized. The image may appear to "jump" when registration is not properly maintained.

Solarization — The partial reversal of an exposed film or print through reexposure during development.

Super — Abbreviation for superimposition. The placing of one image over another so that both may be seen simultaneously. The effect can be achieved in many ways: by more than one exposure on a piece of film in a camera; by glass shots; by double or multiple printing, etc.

Wild Sound — Any sound recording made without synchronous film sequences.

Wipe — An effect used as a transition from one scene to another. It can be used where several projected images form one composite scene. Beginning with one side of the screen, a portion of scene A is replaced (through dissolve) with the corresponding portion of scene B. The effect moves across the screen until only scene B remains. (This term is adapted from the motion picture effect, where one scene replaces another in a similar manner, side to side.)

Work Print — Any picture or sound track print (usually a positive print) intended for use in the editing process to establish (through a series of trial cuttings) the finished version of a movie film. The purpose is to preserve the original print intact (and undamaged) until the cutting points have been established.

The Visual Format Planning Guide

This guide will help you organize the measurements, calculations, and decisions mentioned on page 58.

I. Audience Size

    1. Anticipated total audience _____

    2. Anticipated number of showings _____

    3. Maximum audience size for each showing **(Step 1 divided by Step 2)** _____

II. Room Size

    4. Length _____

    5. Width _____

    6. Height _____

    7. Dimension of overhead obstructions _____

    8. Location of overhead obstructions _____

    9. Maximum screen height **(Step 6 minus 4 feet)** _____

    10. Adjusted screen height **(Step 9 minus amount of interference from overhead obstruction)** _____

III. Actual Screen Height and Projector Placement

    11. Projection distance _____

    12. Type of slides used _____

    13. Focal length of lenses _____

    14. Actual screen height **(from Projection Distance Table)** _____

    15. Sketch layout of screens. Show dimensions and aspect ratio. _____

    16. Screen width _____

IV. Seating

    17. Minimum screen-to-seat distance **(Step 10 times 2)** _____

    18. Maximum screen-to-seat distance **(Step 10 times 8)** _____

    19. Maximum depth of seating area **(Step 18 minus Step 17)** _____

    20. Total width of aisles _____

    21. Maximum width of seating area **(Step 5 minus Step 20)** _____

    22. Total viewing area **(Step 19 times Step 21)** _____

    23. Maximum seating capacity **(Step 22 divided by 4)** _____

    24. Create a room plan based on the foregoing dimensions and calculations. Show seating, screen placement, projection booth. Draw to exact scale. Following instruction on page 68, determine the optimum viewing area. _____

    25. Optimum seating capacity **(Step 23 minus seats eliminated in Step 24)** _____

    26. Is this figure equal to or more than the figure in Step 3?

    If yes, you have no problems.

    If no, consider:

        a. scheduling additional performances.

        b. using a larger screen if possible.

        c. using a different screen surface.

        d. changing the projection distance.

        e. changing the projection setup (front screen/rear screen).

        f. adopting an entirely new visual format.

OBJECTIVES AND BUDGET

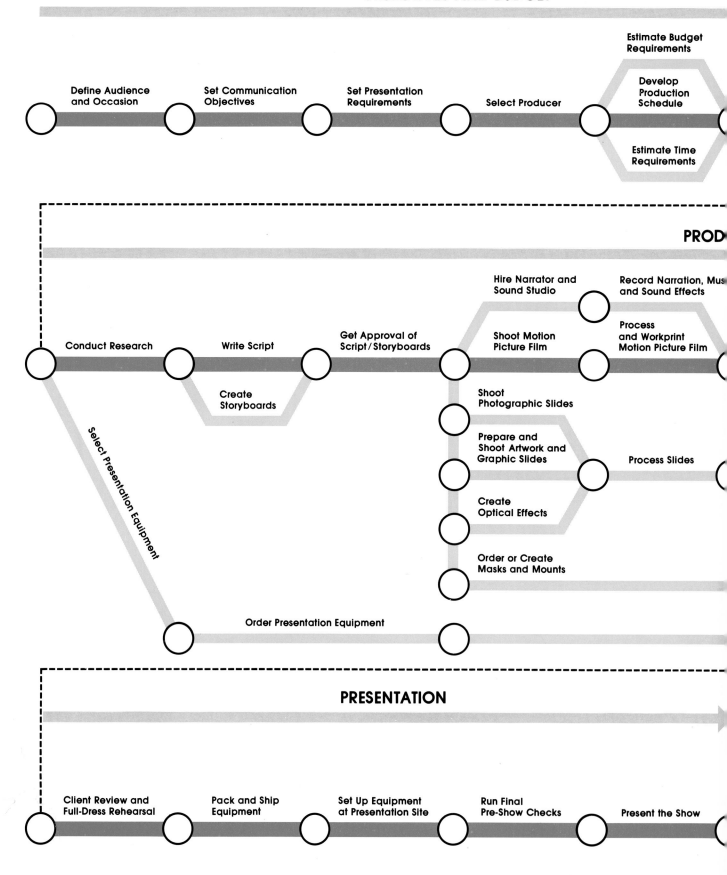

Define Audience and Occasion

Set Communication Objectives

Set Presentation Requirements

Select Producer

Estimate Budget Requirements

Develop Production Schedule

Estimate Time Requirements

PROD·

Conduct Research

Write Script

Create Storyboards

Get Approval of Script / Storyboards

Hire Narrator and Sound Studio

Record Narration, Mus· and Sound Effects

Shoot Motion Picture Film

Process and Workprint Motion Picture Film

Shoot Photographic Slides

Prepare and Shoot Artwork and Graphic Slides

Process Slides

Create Optical Effects

Order or Create Masks and Mounts

Select Presentation Equipment

Order Presentation Equipment

PRESENTATION

Client Review and Full-Dress Rehearsal

Pack and Ship Equipment

Set Up Equipment at Presentation Site

Run Final Pre-Show Checks

Present the Show

Get Budget Approval

Select Media and
Create Visual Format

Create Visual Style

Write Proposal
and Get Approval*

Write Treatment
and Get Approval*

Edit Picture and
Sound Elements

Initial Approval of
Picture and Sound

Conform Motion
Picture Film

Receive and Approve
Answer Prints

Order Release Prints

Get Approval of Programming

Sort and Edit Slides

Initial Approval
of Slides
and Sound Track

Program Presentation

Receive and Test Presentation Equipment**

*We have listed "Write Proposal and Get
Approval" as a single activity because the
two are so closely related. (The same is true
for "Write Treatment and Get Approval.") In
planning, however, you must allow time for the
two activities — you must set a deadline for a
scriptwriter to complete copy **and** for a client,
whether in your organization or outside it, to
read, consider and approve the copy.

**The chart lists "Receive and Test Presentation
Equipment" as the final activity in the
equipment path. That, of course, is an activity
that will take **you** little time to complete;
however, it may take a manufacturer two to
six weeks or more to deliver the equipment to
you. So most of the time indicated on the
chart is to allow for delivery.

How To Use The Chart

The Critical Path chart diagrammed on the inside pages should be used like a road map. When planning a trip from Point A to Point B, you choose a route that suits your purposes best, selecting expressways if your purpose is to arrive at your destination in the shortest possible time, a scenic route if your purpose is to enjoy the sights along the way. In much the same way, when you plan a multi-image presentation, you should move from "Define Audience and Occasion" to "Present the Show" along the path that suits your specific production needs best.

For example, in the production phase of our chart, we have indicated "Write Script" as the critical path activity, the one that will take longest to complete. Your particular presentation, however, may require a great number of detailed storyboards; if that's the case, "Create Storyboards" may be the critical path activity for your presentation.

An area where you are more likely to find our chart diverging from your particular needs is in the planning for motion picture production. We've listed it as a critical path activity. Your presentation, however, may not require motion pictures, so the activities listed for slide production (or, perhaps, sound production) would become your critical path.

The following chart, then, is not a plan to follow, merely a guide to consult. When you sit down to schedule the activities for your production, use our chart as a model, but create your own version of it. List only the activities **you** will undertake, and list them in the sequence **you** will follow.